OCHOPEE

THE STORY OF THE SMALLEST POST OFFICE

as told to
Maria Stone

65th Anniversary Edition

ECITY • PUBLISHING

OCHOPEE
The Story of the Smallest Post Office

© 1989, Maria Stone, original edition
© 2018, Marya Repko, copyright renewed with additional material
   All rights reserved.

set in Times New Roman, 10/12pt
printed & bound in the USA
Fourth Edition, First Printing, August 2018

cover sketch by Wendy Arsenault

ABOUT THE TYPE FACE
"Times New Roman" was designed in 1931 by English typographer Stanley Morrison (1889-1967) for the London newspaper *The Times*.

ISBN 978-0-9830425-7-0

ECITY • PUBLISHING

P O Box 5033
Everglades City, FL, 34139
telephone (239) 695-2905
www.ecity-publishing.com

Other books from this publisher:
   A Brief History of the Everglades City Area
   The Story of Everglades City; A History for Younger Readers
   Historia de Everglades City (Spanish translation by Gloria Gutiérrez)
   A Brief History of the Fakahatchee
   A Brief History of the Smallwood Store in Chokoloskee, Florida
   A Brief History of Sanibel Island
   The Story of Sanibel Island; A History for Younger Readers
   Angel of the Swamp; Deaconess Harriet Bedell in the Everglades
   Grandma of the Glades; A Brief Biography of Marjory Stoneman Douglas
   Memories from Hadlyme; A Personal History of the East Haddam, CT, Area
   Women in the Everglades; Pioneers and Early Environmentalists
   The Story of Barron Collier; A History for Younger Readers
   The Tamiami Trail; A Collection of Stories by Maria Stone
   Everglades Entrepreneur: Barron Gift Collier, Roaring Twenties Tycoon

PREFACE TO THE 2018 EDITION

As we celebrate the 90th Anniversary of the completion of the Tamiami Trail through the glades, it is particularly suitable that this book be republished with its "written oral history" of the people who lived and worked back in that formative era for Southwest Florida.

Author/publisher Maria Stone was a teacher in Immokalee with a love of local history. She interviewed dozens of old-timers to record their memories which she then transcribed into a series of books *(see the list on page 52)* – invaluable resources for other historians.

I am reprinting Maria's original 1989 book (with her revisions in 1992 and 2006) much as it was except for reformatting, adding photos and/or captions, and correcting a few typing mistakes.

My thanks to Lila Zuck, the literary curator of the late Maria Stone's work, for selecting me to republish this fascinating volume about our area history.

If you have comments or corrections or can add more to the story, please let me know.

Marya Repko
Everglades City, FL
mrepko@earthlink.net
August 2018

Note. The original cassette tapes of Maria's interviews have been donated to the Everglades Society for Historic Preservation and will be converted into computer files that can be saved on CDs. See www.evergladeshistorical.org for more information.

Maria Stone ~ OCHOPEE: The Story of the Smallest Post Office

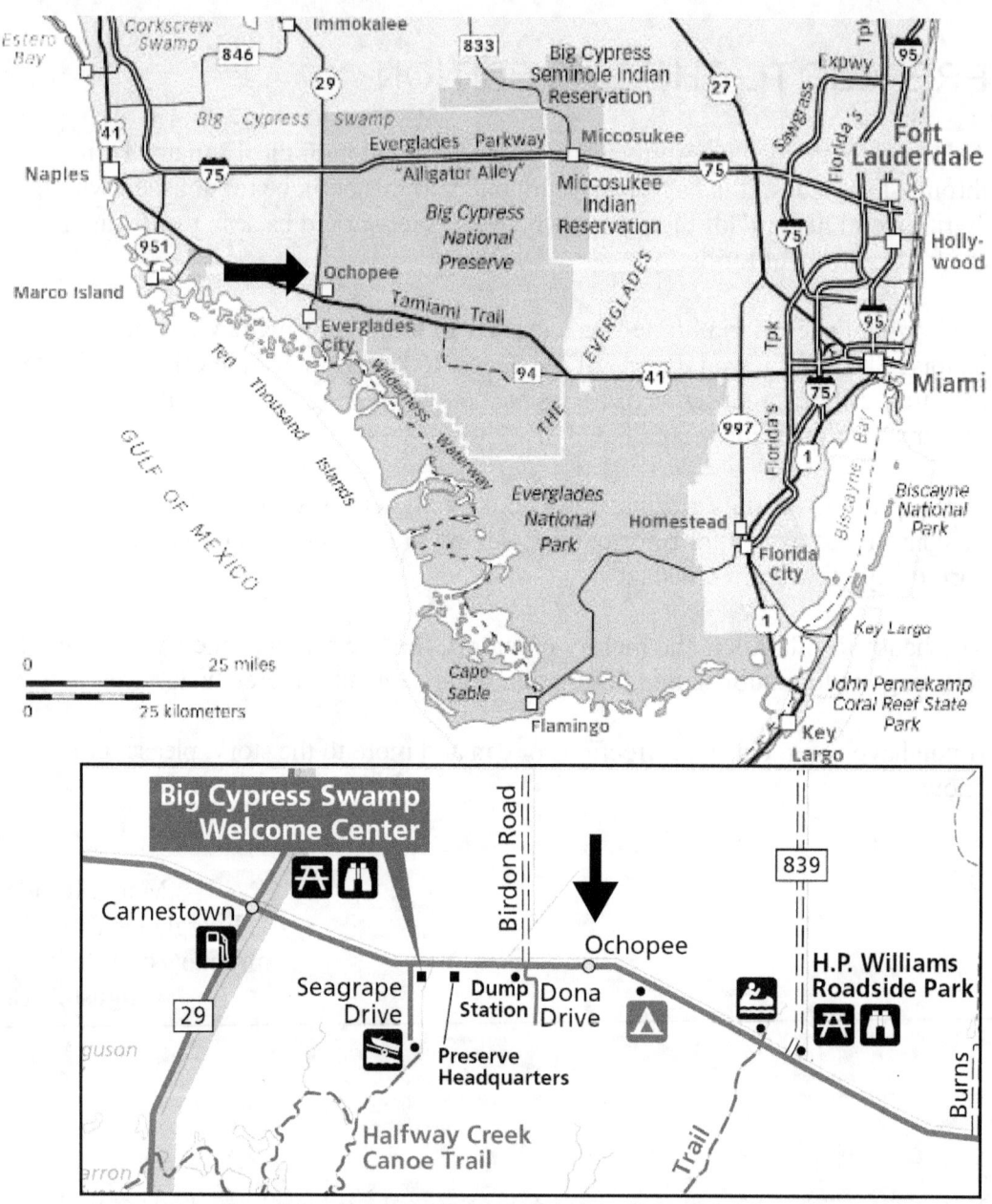

CONTENTS

INTRODUCTION 1
Ochopee and the Sawgrass Song, A Poem by Maria Stone 2

GAUNT FAMILY
Accolades for Mr. Gaunt 5
Buying the Big Field 6
Beginning a New Life 8
Naming Ochopee 10
Quaker Roots 11
Homesteading Miami 12
Growing Up in Miami 15
The Good Life in Ochopee 22
Ochopee Burns 27
The Birth of the Smallest Post Office 29
The Painful End 30
Memories of Ochopee 31
Memories of our Father in Ochopee and the Smallest Post Office 35

FRIENDS & NEIGHBORS
Chief O. B. Osceola Speaks About Mr. Gaunt's Farm 37
Charles Price, Jr: My First Job 38
Col. Frank Tenney: Thank Goodness for the Little Ochopee Post Office ... 39
Bruce Warren: The Mail Route 40
Mr. Meece Ellis: Memories of Mr. Gaunt 41
Lee Hancock: Gaunt's Many Fields 41
Jerri Fish: I Cover the Everglades 42

SHEALY FAMILY & NEIGHBORS
Memorial to Evelyn Shealy 45
Evelyn Shealy: Memories of My Years at the Little Ochopee Post Office .. 46
"Hokie" (Clara McKay): Evelyn Saved My Life 47
Ann Greenwall and Barbara Dwyer: Memories of Our Dear Sister 48
David Shealy, Evelyn's Son: My Mom 48
Vince Doerr, Ochopee Fire Chief: Memories of Evelyn 49

FURTHER READING 51
BOOKS BY MARIA STONE 52
TIME LINE 54

Maria Stone ~ OCHOPEE: The Story of the Smallest Post Office

"**Smallest Post Office Handles More Mail**"—You wouldn't expect Collier County, one of the nation's fastest developing areas, to claim the Ochopee post office as the nation's smallest. There's not even enough room for dead letters. As this small post office does its job, so does United Telephone in handling its communications responsibilities.

Maria Stone ~ OCHOPEE: The Story of the Smallest Post Office

Editor's Note: This tribute and the pages that follow were included in Maria Stone's book. Some of the photos have been added in this edition.

INTRODUCTION

Our history rings out many glorious stories about pioneers who brought forth a community from an impossible dream.

With his own innate Quaker strength, a man by the name of James T. Gaunt and his family brought forth Ochopee. For a time, he did win over the sawgrass. Even now, his spirit must walk the acres of tall grass that were once his fields.

The wind in the grass still sings of the tall, young man with strong shoulders who loved Ochopee and held the grass captive for just a while and then surrendered in the end.

If you take time to listen to the wind in the sawgrass, the tale of little Ochopee will be told to you.

Gone – but not forgotten, because this very little Post Office keeps contact with the outside world.

Mr. James T. Gaunt passed away December 22, 1987.

Postmaster Mrs. Evelyn Shealy says, "A Post Office was established in Ochopee in 1933."

"My postal career began in October 1970 in Ochopee. I was appointed postmaster on August 19, 1972."

Mr. James Gaunt shared his memories with me about his life in Ochopee.

He answered all my questions about the history of this area which he had named and where I have chosen to live.

SYNOPSIS

Day and night, cars whiz along, Miami-bound, on the old Tamiami Trail. Some drivers have an amused smile when they read the sign -- OCHOPEE – which marks no apparent town. A few notice the United States flag flying above a tiny building nestled there in the sawgrass. Fewer yet stop to wonder about its reason for being there by the side of the road.

It is the smallest post-office building in Florida, the smallest post-office building in the United States – if not the smallest Post Office building in the world. With barely enough room to enter, one may hear the pleasant voice of the postmaster, "Good morning. Welcome to the smallest post office!"

She can sell you a picture postcard and a stamp to mail it off to the far corners of the earth.

As the wind whips the flag that's almost too big for the building, you may notice it banging the "Bus Stop" sign as it begs your attention to its presence here in the rippling sawgrass year after year.

The sawgrass whispers the same haunting tune, "I'll win in the end; I'll win in the end!"

ACKNOWLEDGMENTS

The author wishes to thank Daisy Gaunt Brown, Carrie Gaunt Griffin, Dr. James T. Gaunt, Jr., and Betty Gaunt Bryan.

Posthumous credit goes to Mr. James T. Gaunt, Sr., and, of course, Postmaster Evelyn Shealy.

Many thanks, also, to those who contributed stories about life in Ochopee and the smallest post office.

Without their contributions, this book would not have been possible.

OCHOPEE AND THE SAWGRASS SONG

A Poem by Maria Stone

Once there came a man to tame a part of earth,
A home where his children would grow and give birth.
He would tend the land and town he had begun.
'Twas in the Everglades 'neath the burning sun.
Its name was an Indian name – Ochopee.
A town in the Everglades it was to be.

This toil by a man to change the wild – will last.
Then comes the triumphant song in windy blast.
Till and toil, but in the end this all will pass.
"Nothing lasts," is the song of the wild sawgrass.
Day and night it waits; yes, waits there with the wind.
Its song, "Sawgrass back to sawgrass" in the end.

We hear the life of the town was bright but brief.
One night nothing remained but ashes and grief.
Tell the story of the town that burned so fast.
Tell the tale of Ochopee's smoldering past.

They spent its last pennies 'cross that wide expanse.
A weary traveler took a room by chance.
Oh, tell what fate brought the man that night to pause.
The man – all knew – was the culprit and the cause.

Drunk from a bottle, his weed burning his bed,
He ran screaming, nude, flaming from toe to head.
More walls flamed from the traveler's burning bed.
Hungry, raging, wind-fed flames began to spread.
'Twas there on the prairie that night of the fire
The wind's sawgrass song beat flames higher – higher.

Of Ochopee Town not much left to be found.
The night Ochopee burned down – down to the ground
All hands rose when voices rang the "Fire!" shout.
The strong young man, his family round about,
Stood by as their hopes and dreams and years of toil
Rose high to dance with the wind free from the soil.

His loss – oh, how could he pay his men for hire.
Up in smoke his means, his dreams in the night fire.
"Oh, but for awhile the world was mine," he cried.
'Twas the night Ochopee Town burned down and died.

Still stood the man in the light of early dawn.
"My dreams and gains – all I thought I had – they're gone.
"What's that? We must have mail delivered today?
I guess we can – yes, we must; I'll find a way."

He went as the sawgrass sang its windy song
To fields for the shed where water pipes belong.
Today all that's left to prove Ochopee's face
Is the little-shed post office in its place.

The little shed's fame has spread both far and wide –
The little post office by the old roadside.
It did not win, no matter what sawgrass said.
A memorial to the man – here's his shed.

Now, if along that Trail you should choose to go,
Pause at that little shed – Ochopee's P.O.
Behold, by the road, his work, 'twas not for naught.
Think for a moment of the man – James T. Gaunt.

Sidney Brown, postmaster from 1933 to 1971, outside the Post Office.

Maria Stone ~ OCHOPEE; The Story of the Smallest Post Office

fill out and return at o

ESTABLISHMENT
LOCATION OF PROPOSED POST OFFICE

Post Office Department
FIRST ASSISTANT POSTMASTER GENERAL
Washington

IN REPLYING
MENTION INITIALS AND DATE
AF LB

August 24, 1932

FIRST ASST. P. M. GENERAL
RECEIVED
SEP 8 - 1932
DIVISION OF POSTMASTERS

Sir: With reference to the proposed establishment of a post office at the point named below, and in order that the office, if established, may be accurately represented upon the post-route maps, it is requested that you furnish accurately the information called for below and prepare a sketch according to instructions on opposite side of paper, which should be returned to the First Assistant Postmaster General, Division of Postmasters, as soon as possible.

Respectfully,

FIRST ASSISTANT POSTMASTER GENERAL.

#149

Proposed post office, ...Ochopee..., ...Collier..., ...Florida...
(Name.) (County.) (State.)

If the town, village, or site of the post office be known by another name than that of the post office, state that other name here: ....

The post office would be situated in the ...N.W... quarter of section No. ...24..., in Township ...52..S..., Range ...30.E..., of the .... principal meridian, County of ...Collier..., State of ...Florida...

The name of the nearest river is ...Turner..., and the post-office building would be at a distance of ...Two miles... on the ...west... side of it.

The name of the nearest creek is ...Half-Way..., and the post-office building would be at a distance of ...Two miles... on the ...East... side of it.

The name of the nearest office on the same route as this proposed post office is ...Naples... and its distance is ...37... miles, by the traveled road, in a ...N.W... direction from the site of this proposed office.

The name of the nearest office on the same route, on the other side, is ...Miami... and its distance is ...72... miles, in a ...East... direction from the site of this proposed office.

The name of the nearest office not on the same route as this proposed post office is ...Everglades... and its distance is ...7½... miles, by the traveled road, in a ...S.W... direction from the site of this proposed office.

The post-office building would be on the ...East... side of the ...Atlantic Coast Line... Railroad, and at a distance of ...4 miles... from the track. The railroad station name is ...Carnestown...

The post office would be ...Seven miles..., air-line distance, ...South... from the nearest point of my

Signature of Applicant for Postmaster: James F. Jaudon
Date: Sept. 5, 1932

James Franklin Jaudon (1873-1928) was a Miami businessman who advocated the building of the Tamiami Trail and owned land in the area, some of which he sold the Gaunt family. Jaudon was granted a Post Office in 1932 on his nearby farm *(see page 10)*. The facility was moved into the Gaunt store in 1933 and Sidney H. Brown was appointed as postmaster.

GAUNT FAMILY

ACCOLADES FOR MR. GAUNT

Mr. Gaunt was President of the Florida Farm Bureau in 1942. He was vice president and on the Board of Directors for 14 years.

He also served as President of the Collier County Farm Bureau.

Mr. Gaunt was the Vice President of the Lee County Electric Co-op Board of trustees from 1953-1956, and served as president from 1956-1981.

He was on the State Soil Conservation Board and was Chairman of the Collier County Housing Committee. He was a National Rural Electrification Cooperative Association (NRECA) delegate, NRECA member of the Public Relations Committee, and Florida Rural Electrification Cooperative Association Long-Range-Planning Committee member. He was also the Collier County President of the Lions Club.

We remember Mr. Gaunt not only because of Ochopee but also as the first big farmer in this County. He was the first one to import migrant labor, which changed the emphasis from citrus and cattle to farming in Collier County.

James Tate Gaunt as a graduate in 1920 and standing in Ochopee in 1928.

BUYING THE BIG FIELD

James: In 1928, the Tamiami Trail was to be opened. That's a good place to start this story. The Tamiami Trail meant that new land could be settled. Many pairs of eyes were on that new land to be developed along the Trail.

In 1928, I was still single and fancy-free. Though I farmed with my dad, my friends and I were having the times of our lives in Miami. I think my dad decided it was time for me to get settled.

Dad first went out and tested the soil in what was to be Ochopee. He always had the old pioneer spirit. He liked to settle new lands. So, you see, he was the instigator of the Ochopee project, to get me on my own.

Captain Jaudon had a lot of land in Dade as well as Collier Counties. He wanted to sell it to somebody who would develop it. That way, he could more easily sell the surrounding land. Captain Jaudon lived in Miami, so Dad and I made an appointment with him and rode out to see this piece of land that was seven miles from Everglades City. It was also four miles east of Highway 29, which is a crossroads to Carnestown.

So Captain Jaudon, Jim Janes, my dad, and I went out there and rode all around the land. It was just grass – nothing but grass. Sometimes there would be a palmetto head.

Dad said, "This land sure looks good to me."

That day, we bought 250 acres from Captain Jaudon to start, but later bought more.

J. B. Janes, who was along, was going to work there, too.

Some of the land south of the Trail was salty, but the rest was good land. We paid $100 an acre for this land that was to be *Gaunt and Company*, on the new Tamiami Trail.

Dad was excited. As I looked around, I wondered if, in time, a good, new world could come to pass for me out of this grass. I shared my dad's faith, so I believed it was possible, with hard work.

We did not have a thing in the world out there but grass. As far as we could see, there was grass. The first thing we had to do was fix a place for us to live while working.

We went back to Miami and bought some army tents called squad tents from an Army/Navy surplus store. Every night, we were forced to sleep under mosquito nets if we wanted to keep from being eaten alive by the big ones out there. It was April, and, in order to be ready for the planting season, we built the workers' quarters first.

That was the same year that all those people were killed in that hurricane at Lake Okeechobee. Luckily, we didn't get much wind in our new place, but we did get a lot of water. We had water – water, everywhere.

Even before that, we had water on our land. We had been working in knee-deep water most of the time anyway; now it was deeper than ever.

The workers' quarters that we were building were built on pilings. All those pilings had to be driven in that water and the houses built after that. It was a long, hard summer of steady work to get enough housing for our workers by planting season. Working from dawn to dusk, we were able to get them ready for the first season.

Dad was sure the water would recede in the dry season. It did, for the most part, and our land was ready to prepare for planting.

In those days, we used mules, pulling mowers with sickle blades, to cut the grass. When the grass stubble was dry enough to burn, the ground would be as clean as a floor. All this took time and labor.

As early as we could, we cleared about two acres for our seedbeds. The first year, we planted entirely tomatoes. We had about 300 acres to start with. We had to plant our seedbeds in October.

We built up big piles of dirt and leveled it off for the seedbed, because the water got so high the seed bed would have to be high enough to be above the water line. We had a time limit on farming out there, because, from the time we'd plant our seed to gathering time, we were rushed. We didn't have transplants in those days, like they do now.

This land has a high Ph, which runs at least eight. Neutral is around six to seven.

In those days, we had to use stable manure for fertilizer. It had to be hauled in there, of course, by rail. We got some from Tampa, Jacksonville, and South Georgia.

About then, Barron Collier had built the railroad from Immokalee to Everglades City. This railroad ran through Carnestown and opened in 1928, just about the time we needed it.

The manure was shipped in gondolas. We had our men go the four miles to Carnestown and haul the manure from the Tamiami Trail depot to our fields. Getting it in the summertime, we'd compost it in piles about 5' or 6' tall and 25' wide. It had to be turned with pitchforks or the heat would make it white and it lost a lot of its ammonia and nitrogen.

Now, as to the workers, the first year we started our operation, we had a little trouble getting them to come out, because it was a new place. We had to recruit them anywhere we could. After a few of them saw this nice, new place, they told their friends and families about it. More and more came out then. It was a good deal for them, really.

We had a new settlement of houses for the blacks who lived there and worked for us year-round. They were mostly from Miami. We had some black field workers from Miami, but they didn't work out because they didn't get along with the black workers from Georgia. There was trouble between them that would have to be straightened out. Sometimes they had fights at their jukes. It was quite a ways from our house, but we could hear them occasionally. When they held church, their music came across the fields on a warm night. No matter what they did, most of the blacks were back on the job Monday morning after their weekend of reveling.

We had a settlement for the large number of white men workin' there. It was across the canal.

We also had a crew of about 25 Seminole Indians. They were wonderful workers and took good care of the vines. They didn't break them up like some of the others did. The only trouble with the Indians. You pay them off on Saturday and it would be Tuesday or Wednesday before they'd come back to work. They thought they had to spend that money first. They'd spend some of the money for wine; they were bad about that wine.

Corey Osceola's family worked for us. They were good workers and they were lovely people.

We had a real good year the first year. It was a coincidence, but there was a revolution in Mexico just after Christmas. When we started shipping, the market was $2.50 a lug for tomatoes. After the revolution, the Southern Pacific and another railway, the Northern Pacific, shipped some refrigerated cars down to Mexico to pick up tomatoes. The rebellious Mexican burned them up! After that, the railroads said, "No more cars go to Mexico."

That kept the Mexican grocers from getting their tomatoes to our market. Because of that, instead of Gaunt Company getting $2.50 a lug for our tomatoes, we got $5 a lug. We made good money. A little of our acreage didn't produce like it should because it was salty. We had extended a little too far south of the road, but we cleared well over $100,000 that first year.

Now, the wages for our workers ranged about $1 to $2.50 a day. We furnished them a small house and utilities and health insurance free. They lived their daily family lives out there, each according to his culture. There were several hundred in the three settlements.

We people of the Gaunt Company lived in our homes across the Canal and road, which was quite a ways from the settlements.

One of the quarters was east on the south side of the stores and one was west on the north side of the stores. There were three all together. The Georgia Quarters were called Boardwalk, because the water got up so high they couldn't get to their houses without wading water. We built a long boardwalk for them, so that quarters got the name of Boardwalk.

The Indians lived in their own chickees, about a mile from us.

BEGINNING A NEW LIFE

James: In the early 1900s, while I was living in Miami, I ran around with Ralph Brown, who was engaged to my sister, Daisy. He was from Brunswick, Georgia. In about 1922, I went up to Brunswick, Georgia, with Ralph when he went to visit his relatives at Mid River. While we were there, he introduced me to Lois Burgay, the girl I married a few years later.

In 1925, Lois moved to Homestead, Florida. Her uncle and grandmother had a store there, and she came to stay awhile with them. My heart skipped a beat when I heard she was over in Homestead!

My dad had opened up a new prairie land in Perrine on the banks of Biscayne Bay, on Cutler Road. We had a big packinghouse at the farm. My dad and I were very busy farming at Cutler, but once a week I'd ride down to Homestead to see Lois. Sometimes we would go to a dance or just ride around.

Lois had a job playing the organ for the Seminole Theater. That was before canned music, so live musicians were hired to play during the silent movie. I did a lot of waiting around for her, because she played every night. When summer came, Lois went back to Brunswick. While there, she worked in a jewelry store and played the piano at church.

By that time, I guess I sort of knew that she was the girl I wanted to marry. When she left, I missed this lovely, refined, Georgia girl, so I went to Brunswick to try to convince her to come back to Florida with me.

The summer of 1930, I went to her house and walked right in and proposed to her. I was surprised but happy when she said, "Yes!" My new life was about to begin!

Our wedding was performed at the Methodist Church in Brunswick. There was quite a large crowd because she and her family were pretty well known in their town. Lois had gone to college in Macon, Georgia, and had her degree in music from Wesleyan College. She was a city girl for many years when she finished college. That day, I thought about what Ochopee was like, but hoped she could adjust.

By 1930, our Ochopee settlement was quite a thriving town. There was a huge packinghouse, a boarding house, a restaurant, a garage and bulk plant, a large general store with the Post Office in the front, a filling station, and several other buildings. There were the workers' quarters filling a large area, too. Also, there was quite a large settlement of houses across the road and canal from town. This is where the white people connected with the *Gaunt Company* lived, as well as many other families who chose to live there, but none of it was anything like Lois was used to.

For our community, we had our own utility plant. We had a big Fairbanks Diesel system, with another one as a standby unit. By 11 o'clock, it was turned off and we'd use our lamps. It was cozy, but, of course, we were glad when rural electricity became available. I had a nice frame house already built, all ready for Lois. It had four bedrooms, two baths, living room, dining room, kitchen, closets, and porches.

There was a problem, but I figured it would work itself out. The problem was that my friend, Ralph Brown, and my sister, Daisy, had been married sometime back and were living in my house in Ochopee. My father was 87 and still lived in Miami, but he came out to stay with Ralph and Daisy in my house in Ochopee. He had asked Daisy's husband, Ralph Brown, to take his interest and be the manager of the store and bookkeeper for the operation. Ralph became vice president of the *Gaunt Company*. Dad still liked to keep an eye on the project. Also, our uncle, James Brown, lived in my house with them. This was the situation I brought Lois into. I don't think she liked it too well, at first. This city girl felt like she was coming to a jumping-off place, I suspect, but pioneer women were willing to follow their men to the ends of the earth, and Ochopee seemed like the end of the earth, for sure!

Daisy: Well, when James brought his bride to his full house in Ochopee, I figured that she didn't like it too much, but we got along okay. My first child, Mary Elizabeth, was not quite two years old, and I was trying to train her. Yes, I thought Lois had a lot to put up with. Imagine, a young bride coming to her home, with a family of four living there besides an uncle and father. It was a bit much!

In the afternoon, when the children were asleep, Lois and I would sit and embroider.

In the summer, Lois always went back to Brunswick, Georgia, and stayed most of the summer with her family.

When I became pregnant again, I found out Lois was pregnant, too. My husband, Ralph, and my brother, here, decided it was time for us to have our own house. Our house was built next door, exactly like the one James and Lois had. That was much better for both families. James and Lois had a child by then. Their son, James, Jr., was born in 1931. Our daughter, JoAnn, was born in 1932. That was our youngest. James and Lois had a daughter, Betty, in 1932, also.

I always went into my parents' place to await a child's birth. I stayed this time until our house was finished, then returned to Ochopee. While in Miami, I lived in an apartment above my parents' garage with my children.

That's about all I can say, James. You tell the rest.

James: Daisy and I both knew Lois did have quite a time adjusting to life in Ochopee. Finally, she began to make a place for herself in Everglades City. She taught piano lessons to some of the children. It was fortunate for us that we were so close, as life revolved around Everglades City. It was a big social whirl in those days.

Lois and Daisy belonged to the Women's Club and the PTA. The bus took our children to school the seven miles to Everglades. Lois had friends to visit, too. Lots of things were going on, with Barron Collier giving his attention to the area.

There was only one church in Everglades. It was called the Community Church. We had no church in Ochopee yet, but the blacks had a church in their quarters that I had built for them. Many a warm evening, their soul singing wafted across the canals to us as we sat on our porch. Once you've heard the old-time black music, it is never forgotten.

Life was pleasant for us as a family.

James & Lois in the 1940s and their house in Ochopee..

NAMING OCHOPEE

At first, we didn't have a name for our town. People would come along and they'd say, "What do you call this place?"

People from Everglades would come along and they'd say, "You must be crazy, building all those houses and stores and everything out there in that mud and water!"

We really didn't have a name for our settlement until about the beginning of the second or third year, when we were going good. I got the name for it from an Indian boy. There was this Indian that lived about ten miles back in the cypress. He used to trade at our store. He'd come in often and stand around. He was a white-skinned Indian named Charley Tommy. He was pretty sensible. He had some hogs out there in the Everglades and he would come in our store to buy corn for them. He would take that 100-lb. sack of corn and put it on his shoulder and carry it ten miles, to where he had those hogs. He had to be a strong man!

Well, one day, he was in our store buying groceries and I saw him. I went over and asked, "Charley, what is the word for farm, or field, in the Seminole language?"

He answered, "O-Chopp-ee."

I asked him how to spell it. He said, "I don't know."

I decided to name our town that name, but I didn't know how to spell it, either. It should be spelled with two P's instead of one, but I didn't know that then, so I spelled it "Ochopee," and it's been that way ever since. That's how our town got its name.

The "O" stands for big and the rest of the word stands for fields, or farm.

We now had a name that I liked. I did not want to call it Gaunt's Place or Gaunt's Farm. Everybody had just called it Gaunt's Farm. Now that we had a real name, the State Road Department put up a sign and our farm settlement was well on its way to be a real town.

Right west, a mile and a half, of Ochopee, we had what was called Birdon. It was a pretty good-sized place. It was developed after we were through, because they saw our success. You know, farmers are like a bunch of sheep. If a sheep finds a real good place to graze, the rest of the flock comes to it. That's pretty much the way farming is.

We had a very good year and everybody heard about it, so others came to the area.

Now Jaudon was involved, because he had the land, and Bird was a salesman in Miami. So they built a packinghouse *(photo below)* at the place they called Birdon. "Birdon" is a combination of the names Bird and Jaudon. While Jaudon lived in Miami, he was a tax assessor, I believe. H. W. Bird was, I think originally from Colorado. He'd been selling tomatoes for many people in Dade County. He had access to capital if he needed it.

There was another farm, belonging to Frank Roski, but it never did do much.

Our settlement grew and was very successful from 1928 to about 1953. From 1940 to 1960, I was also engaged in farming in Immokalee at the same time as Ochopee.

Packing House Under Construction At CollierMuseums.com

QUAKER ROOTS

My grandfather Gaunt came from England and was a Quaker. He farmed in South Jersey, not far from Philadelphia. I imagine it was about 18 miles out. Grandfather Gaunt had two sons named George Gaunt and Edgar Gaunt. He died in 1912. I remember it well, because my dad went up there and left me to milk the cow.

The house in New Jersey was built by my grandfather, John Gaunt, when he moved from the farm to the little town.

We came from a long line of Quakers. Our grandfather's house still has our great-great-grandfather's clock in the hall. He had the clock during the Revolutionary War. Since they were Quakers, they didn't believe in fighting, so they took the weights off the big clock. They knew they would be called for to make bullets for the war. Our Quaker family hid the weights underneath the manure pile until the war was over. That old clock got its weights back after the war and still ticks off the time today.

The big, four-poster beds are still in the house. The house has two stories and an attic. It was built before 1890, and is still in good condition.

My father's family members lived around pretty close to each other. They all had farms, and were glad to hear of a 17-year-old like me interested in carrying on the tradition. For sometime, I had been fooling around with growing vegetables, like the rest of the family. The summer I was 17, I went up to visit the Gaunt relatives.

On day, my cousin Webber of Mullicahill, New Jersey, asked if I'd like to go over and look at my Aunt Anna's truck patch. We got in his truck and went down to Camden. In those days, there was no bridge over the Delaware River. We got on the ferry to cross. We were about the third vehicle from the end of the ferry. The fellow that ran the ferry must have been inexperienced, because, when we went in, he hit the piling and everything shifted on the ferry. It broke the chain, and that third truck behind us dropped off the end of the ferry and hung there, half on the ferry. He backed up the ferry and came in again toward the landing platform. He came in fast and so hard that he crushed the front wheels of that truck. The men at the dock finally had to get a block and tackle and pull that truck out.

We finally got off that thing and, believe me, I was glad. That was enough excitement for this 17-year-old fellow from the quiet little town of Miami.

My relatives took me over to the big farmers' wholesale market in Philadelphia. It was a sight that excited me, for sure. I guess they knew it would. When a big pack of vegetables or fruit was brought up, the auctioneer started his chant. The representatives for retail stores and such bid on the produce for sale.

I was impressed with the prices that the farmers received. That cinched it; I was now seriously ready to grow vegetables on a large scale.

I was also meeting my relatives and learning about the roots of our family. They made sure I saw my dad's high school, which was a Quaker school called "Friends' School." They also showed me the house where my dad had lived.

I came back to Miami inspired and ready to grow vegetables like my dad.

HOMESTEADING MIAMI

The main reason I'm including some of these details in this part of the story is so that you can understand just how uninhabited and primitive these areas of Florida was only a few years back. When I go to Miami today, I can't believe I'm in the same place until I go by our old homestead on S.W. Eighth Street.

When my dad came to Miami, two families and a bachelor lived there. He left New Jersey at 19, when his mother died, and began winding his way to Florida.

He stayed in Clermont for two years. He did a little fruit-and-vegetable farming with a friend from New Jersey. He worked in the packinghouse, too. In those days, the growers didn't have the grading and sizing machines, so all that had to be done by hand.

Dad's partner had a brother in New Jersey who wanted to come down to Florida, too. They planned for him to keep house and do the cooking for them. When this fellow arrived, he was just too dirty to have around the kitchen. My dad said he chewed tobacco and let it run out of the corners of his mouth and down his shirt.

My dad couldn't stand that, so he told his partner, "I'm going to Biscayne country."

He caught the train and rode to the East Coast and went on down to Palm Beach.

Now, Mr. Flagler was just building the big hotel in West Palm Beach. He had to bring his workmen down by boat from a little town north of Palm Beach. You see, the railroad hadn't yet reached Palm Beach.*

Dad said he heard some of Flagler's friends say, "You must be crazy, building all these hotels down here."

Mr. Flagler answered, "Well, I intend to have a lot of tourists down here and they have to have some place to stay."

Now, the first hotel was in St. Augustine. Then he built one right next to the beach going to Daytona. He built some others right on down the coast, also. After Palm Beach, he built one in Miami and one in Key West.

My dad got a job working for Mr. Flagler in Palm Beach. He noticed that most of the men hiring out called themselves first-class carpenters. All they had was a hammer and saw. Dad said that, where he came from, in New Jersey, men were really first-class carpenters, not in name only.

Well, Dad did not want to misrepresent himself just because he had a hammer and saw in his hand, so he hired out as a laborer for about a month, before moving on. This gave him eating money.

Now, up north of Palm Beach, the railroad terminated. A barge then brought all the supplies for building the hotel on down to Palm Beach. From the railroad terminal down south, there were no good, hard roads, either. They built a boardwalk from Lake Worth over to the shore, where the hotel-construction site was. Men would unload the barge and put the lumber and other supplies on a pushcart, which was then taken across the bridge to the carpenters.

Now, the lumber had to be piled in a certain pattern and just right, so that it wouldn't topple over as it was carted across the bridge. Well, stacking this lumber was the job my dad got.

When he came to work one morning, he was busy trying to correct some other man's mistake in his stack. Just about that time, the big boss came along and the whole pile fell over, right in front of him.

He got on Dad, hot and heavy. Dad was put out about the fact that he wasn't allowed to explain that it really wasn't his fault. Dad ended up telling him that he was quitting, because he resented the accusation. Anyway, my dad knew he was meant for more than stacking lumber.

---

* *Editor's Note:* Flagler's Florida East Coast railroad reached Palm Beach in 1894.

Dad took a coach that day down to Fort Lauderdale. He spent the night there. Nearby was a big tent, like a gospel revival tent. In this tent, the cooks served food, family style, on long tables. It was cheap, too. There was plenty of native meat to eat. (In early days, people got their meat from the Indians nearby. It was venison, wild turkey, and fish.)

This lady who ran this tent just put everything on the tables and a man could eat all he wanted for the price. Dad always said that was the best food he ever ate in his life. I guess he was pretty hungry.

The only mode of communication was the mail, carried by a sailboat, which made two or three trips a week along the coast.

Dad wanted to head on south. He waited two or three days for the mail boat to come along. Finally he got somebody to take him across the New River. There was no bridge across it then.

An interesting little Indian story is about how it got its name. According to geologists, that river appeared overnight. The Indians say this is exactly the way it happened, so they named it "New River."

Evidently, what really happened was that the river was underground but close to the surface all along. Suddenly, the layer of soil over the underground river just caved in.

Well, when Dad got across the New River that day, he walked on down to Miami. That is 24 miles. He made it before dark to Miami. My dad could really go out and walk when he had a mind to.

In Miami, he met up with the very few old-timers that were in the area. One that he knew real well was named Davis. He'd been around a long while before my dad got to Miami. Davis knew all that Miami country. Where Miami is now was just a hammock in those days.

Now, Mrs. Tuttle had a little store down on the river. Of course, everybody traded with her by the method of barter for the goods that she had in her store.

In the countryside, some of the old settlers had sawmills. Coontie (roots of wild plants in Florida) mills were also common. Some of the old-timers would go out in the woods and dig up these starchy coontie roots with a grubbing hoe. That root didn't grow in low places, because it would rot from too much moisture. It grew on high ground. The workers would load up these roots and haul them to the mill. The coontie mills looked very much like sugar-cane mills.

There would be a long pole fastened to the mule that walked around and around. Yes, it was very much like a cane mill.

The grinder was a log with spikes driven in it. As the mule walked around, the grinder turned, breaking up the coontie roots. After the roots were ground up, they were put in a barrel and water was poured over the pulp and it was set to ferment. During this time, the pulp would float to the top and the starch would go to the bottom. After awhile, they'd pour the top solution off. It would be brown, and, supposedly, that part of the coontie is poison, but I don't recall how they got rid of the poison.

The starch from the coontie settled to the bottom of the barrels. It was washed good and put to dry in the sun. It was then taken to the store and traded for whatever was needed.

During those first few days my dad was in the Miami hammock, he got a job in the countryside to dig for coontie roots. This helped him get by.

One day, one of the men in the hammock said, "You should get you some land. There is a section of 160 acres on the opposite corner from the coontie mill that you could take over."

The man who homesteaded the land on which the coontie mill stood was Fisher. The northeast corner was homesteaded by a man named King. The southeast corner was homesteaded by Belcher. The man who owned the other corner was a purser on a steamship line. He only had a little shack with a palmetto-fan roof on the land. There was a little, low door, just big enough to crawl inside. He only came to Florida about once a year. He never improved the land in any way when he came.

In early days, the Homestead Law stated that a claimant must do so many improvements on the homestead each year. It was evident that the purser wasn't doing that. The other homesteaders by him were very active men. They became important businessmen in Florida – you know of that.

Well, my dad took his friend's advice and contested the purser's claim. There was no trouble in the contest, because my dad built a nice log house on the land. He did some farming as well as making many other improvements. He planted some oranges and grapefruit, too.

This meant all four homesteads were being developed. This was Miami.

About that time, Mr. King's fiancée came down from Georgia to marry him. Before long, Mrs. King's 15-year-old sister came down to live with Mr. and Mrs. King. My dad got to going with that girl, Mary Alice, in 1897. When she was 17, she married my dad. Before she married him, though, she laid down some conditions he had to meet.

When Dad was dating Mother and finally asked her to marry him, she said she wouldn't marry him until he built her a house, and she didn't want any log house. He built an addition of lumber onto the log house already there. He covered the log house with lumber and it became the kitchen. Since pine land isn't the best land to farm, he then bought some land west of Miami – right on the edge of the Everglades – which was sand land. He became what was called a truck farmer.

Dad built a house out there, too. That was S.W. Eighth Street. It was a good location for him, as the Seaboard Railroad was on one track and the Florida East Coast Railroad track was parallel to it. This meant he had good transportation for his produce.

Before the Seaboard Railroad came into Miami, the Florida East Coast Railroad was the only one to serve the Miami area. When the Seaboard built their tracks, they bought 200 feet of Dad's land, which had bearing grapefruit trees on it. Before he sold the land, he dug the trees up and moved them to another plot and saved them. He grew beautiful potatoes, eggplant, and other vegetables and watermelon. He had a way with plants, just like his Quaker ancestors in New Jersey.

In 1917, there was a hard freeze, which froze his trees back about six or eight inches – but they survived, with Dad's expert care.

Then, S.W. Eighth Street was just a trail through the woods. Different settlers along the way kept improving it. The men of the area made a road by going out in the woods and getting surface boulders and laying them down in the road and breaking off rough edges so there would be a roadbed.

This place on S.W. Eighth Street was eight miles from the original homestead. He built a new house on that land in 1926. Years later, he sold the original homestead, which is now called Bryon Park, in Miami. After that, he bought some property on Cutler Ridge, which was 200 acres. That was good farmland. He did well there.

I was born in the homestead in 1900 (53 years before the little Post Office). My dad had lived on that homestead spot since 1896. I had three sisters. The oldest one, Daisy here, married my friend, Ralph Brown, a Georgia boy. My middle sister, Carrie, lives in Sunrise, Florida. Daisy, Carrie, and I were all three years apart. My youngest sister, Edith, was born nine years after Carrie. She married David C. Brown, the Collier County Commissioner, and lives in Immokalee.

It is a coincidence, but David Brown and Ralph Brown are not related.

As a little boy, I remember several instances I can relate about life in Miami back then. Dad was always going out to the farm to oversee it. Lots of times, I ran down the road to meet him. Sometimes he'd ride a bicycle and sometimes he'd take a horse and wagon.

One particular time I remember Dad took this mule that was very skittish. He was driving her on 22nd Avenue, about a mile and a half from the Eighth Street place. This mule was very, very scared of Indians. She could smell one a mile from her, and when she did, she would run away. Of course, other things could make her run away, too. This one day, a piece of paper flew up and she jumped and broke the singletree, but the breeching held on and so she was still pulling the wagon. She got to running so fast that Dad crawled to the back and got off.

Now my mother and I were waiting for him at the gate. This old mule ran right on by us, without my dad in the wagon. We had a driveway that ran under an elevated water tank. This mule went under that tank and never touched it. She ran under the shed and there she hit a pole that broke the breeching. The wagon dropped and that stopped her.

As you can tell by this, Miami was very "country" in those days!

Dad helped dig out palmettos and rocks on 22nd Avenue, so the few people there could get their wagons through. Dad helped build the little one-room schoolhouse where we went to school. Later, it had two more rooms added. I remember that he gave land for a community church also. He did a lot to help the community develop.

We could roam over that 160 acres of homestead and the new place on S.W. Eighth Street as well. That's where we grew up; we went to school in the little schoolhouse Dad helped build, and went to the church he gave land for. We felt like we really belonged.

My sisters helped in the house and gathered eggs and learned to sew.

Life was simple but good in Miami, in days gone by.

GROWING UP IN MIAMI

Well, my boyhood days in Miami weren't as calm as my sisters' days. I was out looking for adventure with my friends when I wasn't picking beans or doing other small jobs on Dad's farm. Mother had a wood stove to cook on, and it was my job to bring in the wood for the stove.

I decided the easiest way to get that wood in was to drag it in. I found a piece of bailing wire that was fastened together. I put in on the chopping block and picked up the axe. I put my foot on that block to hold the wire. I missed, of course, and cut my big toe about off!

I screamed and the blood poured. The toe was cut through the joint, and it was hanging by the tough part of the skin. I started running to find Mother. Every time my heart would beat, the blood would squirt as I ran through the house.

A neighbor came along in his buggy. He hurriedly called the doctor from somewhere. Dr. James M. Jackson came in his Maxwell car. He laid me on the couch and he sewed that toe back on. Dr. Jackson said it was remarkable how that toe healed.

Kids can always find some things can be done another way and get into trouble!

Miami River had a fork in it. They dug a canal all the way into Okeechobee. The other fork was just a creek about eight feet deep. My friends and I used to swim there, near Musa Isle. We'd pull all of our clothes off and leave them on the bank. If one of the boys had to go home early, he'd tie all our clothes in knots.

When we were small, we climbed trees. West of the house was a clearing, and pine sapling had grown up. We boys would go over and climb the saplings. They were so close together that we could get in the top of one sapling and give a big swing and go on over to the next sapling.

One time, one of the boys was in a pine sapling and the top popped out and he fell. There was a rock under the pine needles and it shook him up pretty bad. We didn't stop swinging in saplings, though.

In 1910, Dad's father was real sick in New Jersey, so Dad went up there. We had this cow to milk. Grandmother came over and we were all in a stew about milking this cow. Grandmother had milked the old range cows when she lived in Georgia. Her way of milking was to milk with one hand holding a tin cup and pour it in the bucket.

Dad always milked our cow with both hands. I hadn't milked but just a little when Grandmother said, "I'll do it."

Between Grandmother and me, we got her milked. I managed to learn to milk that cow before Dad came home.

Later, when I was a little older and I had to milk them all, the neighbor boys and I would get to playing and it would be dark. I'd still have to go hunt those cows and milk them. I put a bell on one so I could find them in the dark.

Soon, we moved out to the other place. The two-story house on S.W. Eighth Street—it's still there, by the railroad, today.

Mother was critical if my grades weren't so good. Dad didn't seem to care too much. I walked to school with my cousins, who were children of my mother's sister. Their names were King. We walked about a mile to our schoolhouse.

I guess even at that early age I knew I wanted to be a farmer like my dad. I liked growing plants. I knew that I didn't want to be a dairyman, after having to look after those cows.

Dad had two places on the Trail, which is Eighth Street. The original house had two stories. He built a new house on the other land. He had horses and mules, because we had no tractors to work our fields back then.

Speaking about what kids will do—one time, when we were living on Eighth Street, on the Trail, a cousin of mine, Gene, lived in town but we got together often. Back of my uncle's house, someone had planted a big watermelon patch. Gene and I got three or four other boys to come out to steal melons.

I had some firecrackers. Gene said, "Now, when I get those boys in the patch, you throw one of those firecrackers down."

I was hidden there behind the fence in the Australian pine trees. Those boys were out there, thumping melons, having a great time. When they got real busy, I threw one of those firecrackers down. "Boom!" went the firecracker. Some of those boys fell on the ground. They thought they were shot. Some took off to the woods.

That night, Gene and I had to hold each other up we were laughing so hard. We told the boys later that Gene and I were playing a trick. One of those kids never would believe it was a joke. I still think it's the funniest yet.

Growing up in Miami in my time was entirely different than today. It was a slow-paced, peaceful place to live, back then. We kids swam in Biscayne Bay, as well as the Miami River. There was no pollution in them. With not many people around, there was less chance of water being contaminated.

Dad would drive the horse and wagon right out in the stream or lakes and fill up the barrels for the house or for produce. It was all pure water, those times.

You know how kids go poking around empty buildings. Well, we found some great places to have fun.

One thing we boys used to do was take our motorcycles out to this old racetrack we found. It was about a mile or two behind the high school. This track was built for bicycle racing. It was oblong, and the walls were about 20 feet high and made of wood. The place was never locked. We could roll in there and go at it. It was like a motor dome. You have a little catwalk to get started, then, when a rider gets speed up, he goes up on that 20-foot wall.

I got pretty good at it. The end walls were perpendicular, but the sidewalls were about a 45-degree angle. When we came down off that straight wall, we had to be not over halfway up that wall, or we'd hit a rail at the top, which kept us from going out. If we hit that rail, we'd lose control and then we'd slide down the wall.

I found out about the second time I went around that thing that, when I got to the perpendicular wall, I should be about three to four feet from the catwalk. When the wall changes angle and the momentum of the driver changes direction, that gives plenty of room to straighten that thing out.

One kid didn't realize this, and he hit that top rail. He got all skinned up as he came falling down that wall. He was just lucky. He could have been hurt pretty bad.

Most of us developed skill enough that we didn't go home all skinned up. Our parents never knew what we were doing. They didn't think they had to worry about us boys riding our motorcycles around Miami.

Those days are wonderful memories!

I started riding a motorcycle when I was almost 15 years old. I had two or three different ones. I drove one to high school every day. Part of the time, my sister, Daisy, would ride on the tandem.

During the summer that year of 1917, I decided I wanted to go up to New Jersey to visit relatives and work on their farm. I put my suitcase on the back of my motorcycle and strapped it down. I had a time keeping it on there.

I was riding a Harley Davidson, and you get a lot of bounce from them. I just couldn't keep that suitcase back there. I think I raided two barns on the way, hunting baling wire to tie it down.

The first day, I rode from Miami to St. Augustine. That was a pretty good trip. The roads in those days were not so good. They were only nine feet wide. They were not only narrow, but they were brick.

I spent the night in St. Augustine. The next day, I reached South Georgia. I spent about a week with my sister, Daisy, and her relatives. Then I went over and spent a couple of days in Brunswick, Georgia, with Daisy's sister-in-law. There were no roads from Brunswick to Butler Island Plantation to Darien. The only way one could go was to ride on a flatcar from Brunswick to Darien, Georgia.

Someone told me, "Well, you can go up by river steamer if you don't want to ride a flatcar."

I was plenty glad to hear that. I went down and made arrangements for passage for me and my motorcycle. That day, the tide was about 15 or 20 feet above the dock so we went around and went up the Inland Waterway. We went up Altamaha River.

At the dock, we arrived at high tide. It was a good ways down to where the motorcycle was – on the top deck – down to the dock. One of the hands says, "We'll just pick it up with our crane and put it down there for you."

I didn't want to take the chance of that motorcycle being damaged, so I said, "Thanks, but I'll just ride it."

So I rode that motorcycle right on down that gangplank.

I went to Savannah, Georgia. Most of the roads then were made out of shell. When I got to Savannah, I went to one of those little railroad rooming houses and parked the motorcycle in the back yard. I was hungry, so I went in to get something to eat.

I went to bed immediately, as I was pretty tired. Now, most of those houses then had no screens on them. I expected mosquitoes. Pretty soon, something started to bite me. I was used to sand gnats and I thought that was what it was. I pulled the sheet clear over my head, thinking they couldn't get to me. They were still biting me pretty bad. I got up and turned the light on and there were these things crawling all over everywhere. Big bedbugs!

I went downstairs and I told the lady, "You know, I can't sleep in that bed where there's bedbugs!"

She says, "Well, son, here's your money back. Y'all just go on, then."

I went out and fired up my motorcycle, as tired as I was. I rode to this ten-story hotel and finally got a little nap.

Next day, I got on my motorcycle and drove up almost to Augusta, Georgia. All that day, it rained and rained. Those highways were slick as could be.

I'd be riding along and that motorcycle would slide right out in front of me. I picked that thing up five different times. It was heavy, too. One time, I slid around onto the grass shoulder. It happened to be drier, so I rode on the shoulder to the next little town.

I'd had enough. I saw this garage and I went in and asked this fellow, "Sir, would you store this motorcycle for me?"

He said, "Yes. Put it over there in the corner."

I went downtown where there was a rooming house and I went in and asked for a room. I told the lady, "I had an experience with a bedbug-room last night. Have you got any bedbugs?"

She says, "Oh, my, no!"

As she took me up to show me the room, there were twin beds in the room. She says, "Now, there's a salesman that comes in town. He may come in here tonight, so don't let that bother you."

As soon as I saw this room, I thought to myself, "I'll bet it's the same darn thing." Well, I went to bed and it wasn't long until those bugs were all over everywhere.

I was ready for a good night's rest after that hard day of rain and sliding around on that motorcycle. I got up and started killing bugs. I got all the big ones killed and the little ones came out. It was about two o'clock in the morning by then.

I took my suitcase and went out on the front porch. There were some real good rocking chairs, so I spent the night in one of those. When the sun came up, I walked down to a restaurant and got some breakfast. From there, I went to the depot and bought myself a ticket to New Jersey. I'd had enough!

I went to my grandfather's place where I stayed two months.

Another thing we fellows did in Miami was to search for treasure. This friend of mine, Charley Brookfield, and I owned a 32-foot cabin boat together. When we weren't riding our motorcycles or doing some other thing, we were in that boat. We let some friends take our boat up to Choctahatchee Bay. That's near Ft. Walton. They were to stake out the place where some Spanish treasure was buried. Charley and I were go to up a little later.

A fellow we knew in North Hollywood, Florida, had an old parchment paper that showed where the old Spanish treasure was hidden. The boys took our boat up there and some buoys to mark the place. They got in some problems. They didn't get much accomplished because the man who owned the land ran them off with a shotgun. Those boys left our boat and got a ride back to Miami. I think they ran out of money to buy the gasoline for the boat so they left our boat up there. Charley and I had to go after it.

A fellow carried Charley and me over to our boat. We were relieved when we saw it was okay. We gassed it up and went across the bay to this bayou. There is a pass called Leach Pass. It goes at an angle, and a boatman can go by and not even see it. We found it all right.

Charley had gotten acquainted with this man where our boat was moored, so we borrowed a bateau (rowboat) from him. We paddled round in that thing until we found the place that the treasure was supposed to be according to the parchment chart.

According to the chart, we had to set the range right on the shore. There were to be oaks. They were there, all right. The beach was just sand.

We got overboard the bateau to hang lanterns out. The idea is to line them up by a certain degree, then, when you go out in the boat a certain number of feet from shore, that's where this treasure was supposed to be.

We had a big ball of cotton string, and we marked the spot and anchored the boat. We were doing this all in the dark, because we didn't want that fellow to shoot at us the way he shot at our friends.

On our way to Choctahatchee, we had bought a bunch of fishing poles. We used these to sound all around where the boat was anchored. With one, we hit a place where the pole wouldn't go down, so we marked that place. Those poles stuck up all around in the water.

Late that night, we went back in to shore and went to bed. The next morning, we got up before daylight. Since I was the best swimmer, it was up to me to go out and investigate the treasure. I swam around all of the places we had worked so hard to mark.

Well, our "treasure" turned out to be nothing in the world but beds of old, dead oyster shells! After studying that map so carefully and doing all that tedious work, that's what we found – dead oyster shells.

After I reported my find to Charley, we sat awhile, letting our disappointment die down. We had been so sure of treasure. We finally got around to taking the bateau back to the man. That was the end of our treasure hunt.

I guess young fellows need that kind of adventure in their young lives. As I look back, the anticipation and the preparation were worth it all.

The next day, we took out eastward in our boat to Panama City, which was about 25 or 30 miles. On the way down, we were fairly close to shore. Soon we saw all these cattle offshore, with their heads sticking out of the water. About that time, we found out why those cows were out there. Those darn blowflies were swarming all over us. We had to go way out from the shore to get away from those things.

About dark, we decided to go into the dock at Panama City. I've never seen anything like it ever again. That water was working alive with phosphorous. The fish were thick on each side of our boat, playing in that phosphorus. It was a sight to see.

We stayed a couple of days in Panama City. From Panama City to Appalachicola River, we took the Inland Waterway. We didn't fool around with those capes, because we knew they would mean bad weather, as a rule.

At Appalachicola, we went into the dock, as we had to have a new shaft put in the boat. While there, we happened to go into the Post Office and we saw this sign about a hurricane, but we weren't too worried. We got our boat fixed, and the next day we took out. We didn't even have a barometer in the boat.

There was an island near, called Dog Island. We debated about what to do, and decided to cut all the way across that island to Tampa.

The waves by that time were pretty high. Finally, we went up to St. Lawrence Island. Before we got there, we got in that storm proper. Waves were coming up over that boat.

You know, the deck of a boat is supposed to be watertight but so much water came over the deck that it was filling up the bilge. We didn't have an electric pump. I just had one of those simple little hand pumps.

I guess I was just too excited or in too big a hurry, but, before I knew it, that thing went overboard. I had on some rubber boots, so I just reached down and cut me a new piece of rubber out of my boots and put on the lever.

That old engine was very reliable; it had two spark plugs for every cylinder. One was on the battery and one on the magneto, so it had dual ignition.

Just to show you how rough it was, the compass flipped clear over. We couldn't tell where we were going with that compass doing that, so we picked out a star. I got one of my poles and, every little bit, I'd take a sounding. At last, I hit bottom, and I said, "Charley, this is where we are going to stay!"

I threw the anchor over and we stayed there until the next day. You'll get a certain drift to an anchor of a boat, but we got along. By the time it was almost dark, we saw a fishing boat out aways. He was bobbing up and down. We got close enough to talk to him. We asked him if he were going in pretty soon.

He said, "Yeah. You boys need help?"

We told him we did. We hooked onto his boat and it was pitch dark, but we went right on in with him.

We got out on the dock and went to get something to eat.

The next day, we went to a spot just off Tampa and stayed the night. The next day, we made Sarasota.

Charley and I would take turns at the wheel. We left Sarasota the following day and went to Boca Grande. We got almost down to Captiva and we tied up on a little mangrove island. The water was four feet deep when we anchored. The next morning when we got up, the wind was bringing in breakers and was thrashing us about.

The wind kept getting stronger and stronger. Charley wanted to go on down to Sanibel, because we were about out of groceries. I told Charley, "I believe that doggone thing is a hurricane!"

We had on 80 pounds of anchor, and yet the wind blew us around. By the afternoon it quit. That was the eye of the hurricane passing over.

Charley said, "Let's go on down."

You see, he had never been in a hurricane. I was native to Florida, and I knew it was a hurricane. I said, "Charley, I think we should stay right here."

We went around on the other side of the island so the wind wouldn't blow us offshore. There were ten feet of water on that side, and the bottom was oyster shell. We had to cast that anchor about five or six times to get that 80-pound anchor to hold.

The first time I went out to throw the anchor that wind and that rain drove me back in the cabin quick, but I knew I had to keep on until I got that anchor out.

After that, we went on two-hour shifts. We spent the night on that mangrove island.

The next morning, we went on into Sanibel. The old Mathews store was located out on the dock. The piling had washed away and water was six feet deep in the store. They didn't have any food except canned stuff, so we got that. They told us, "Captain McKenzie and his ferry boat will be here in a few minutes, if you boys will wait."

We waited and got some bread. We went on down to Marco.

When we got to Marco Pass, we got into that ground swell that they have, especially after a hurricane. That boat would have turned over, but Charley kept her straight on course. We spent the night on Marco at Grits Griffin's store dock.

The next day, we went on around and went out by Coon's Point, out from Goodland, on around out to Cape Sable and back up by Coconut Grove. When we got to Coconut Grove, we looked up there and saw all these little boats blown around everywhere. Everything was all blown down – even the markers.

I told Charley, "This thing must have really blown here!"

We were gone about two weeks on this trip. Neither of our parents knew where we were. They knew there was a hurricane, but they didn't know where we were in relation to it.

That was the 1926 hurricane, and our treasure hunt!

I went to Cuba more than once, but I remember one trip that was memorable. I had a close friend in Miami whose father was also in the fruit business. In Miami, there was a good market for avocados. Dick's father had a young avocado grove in the countryside, out of Havana, Cuba. These seedling trees were about 30' high. The Cuban workers would gather them by shaking the trees and letting them fall to the ground. When the avocados hit the ground, they would be bruised or cracked open.

The Brooks Fruit Company wanted every avocado for the good market in Miami, so Mr. Brooks told Dick and me to go down to Cuba and impress upon those Cubans that they were to handle those avocados like eggs.

We rode the train to Key West, then got on a boat to Havana.

There was a fellow from Belgium in Havana selling heavy sugar-cane equipment. His sideline was looking after this fruit business for Brooks. He was to take us in his Stutz to the avocado fields, about 30 miles out of town.

His Stutz roadster was used to race on the racetrack in Havana. He had a racing body that he put on when he wanted to race. It was a honey of a car.

Well, he got us in that Stutz and he drove that through all those little villages at a racing speed, with the chickens and the kids running all directions. The dogs were barking at us, too, as we fairly flew out there to that avocado farm.

I guess we got the avocado problem solved, because the avocado shipments were large after that. While Dick and I were there, we had a whale of a good time, as young fellows would have.

I found out I had to be awful careful of what I ate. Dick got the dysentery first. He had fever and was out of his head, talking crazy. When he got all right, then I got it. I was ready to come home, but Dick says, "No, let's stay a whole month."

We did stay long enough to see the sights. That country is lush and beautiful. Dick still has a big packinghouse down there; I'll bet he remembers the fun we had in Cuba. I know I do.

Yes, looking back, I know that I grew up in Miami in "the good old days"!

Ralph & Daisy Brown family in 1938 (top). Sidney Brown family in 1954 (bottom).

THE GOOD LIFE IN OCHOPEE

James: Now let me get back to the story you came to hear.

Ochopee was well established. Things were going along smoothly. We had some years that were better than others, but life was good. I, like my family from way back, seemed to realize that, when you are a farmer, you have to keep on trying. Eventually, you come out.

We had quite a few people living in Ochopee. Ralph's brother, Sidney Brown, and his family were there. He was a clerk in our store and also our first postmaster when we got a Post Office in the front of the store. Before that, we had to go to Everglades City for our mail every day.

There were my wife and me and our two children, Daisy and Ralph and their children, and Raymond Cail and his family.

Raymond was a good friend of ours when we had lived in Miami. He farmed in the vicinity of Ochopee and used our packinghouse.

There were also J. B. Adams and his wife. The Vandivers were there. Macadoo Littlefield and his wife came down from Georgia. They had two cute little girls. Ralph called them the Gold Dust Twins. The Griffins had two boys. The Vandivers had a son. It was a close-knit community.

I guess I'll let Daisy tell you about what life was like for the women.

Daisy: Well, I'll say life was good out in Ochopee. We all had children about the same age. They played around and found their own entertainment. They weren't allowed to go to Everglades, seven miles away, unless they went with their parents.

We women looked after our homes and our children. We belonged to the Women's Club in Everglades. We went to church on Sunday. We'd visit with each other.

Ralph and Mr. Cail built a barbecue pit in our back yard. Ralph, Sydney, and Mr. Cail loved to go hunting. We had turkey and venison. Ralph had a boat, so he went fishing. We had a big iron pot that we cooked the fish in. Also, we made venison stew in it. When we had a cookout, every family brought a covered dish.

Our most fun thing to do was to go to Marco to swim at the beach. In those days, cows ran the beach and we had to watch where we stepped. We'd take our children and some of the others. We would fish and pick up shells. We would build a fire and cook on the beach. My husband would dig clams and we'd fry them, too.

Those were wonderful times.

Joe McKay lived east of us a little piece. He worked at our filling station and our bulk plant. He had some bears down there in a cage. Also, he had two or three deer in a fenced area. It was permissible in those days to keep animals like that.

Mosquitoes were bad, but worse in Everglades City. We always had to fight them in either place.

There were some alligators in the canal. An otter now and then. An otter is a lot of fun.

There was little social life for us. We families really didn't get together at night unless somebody had a birthday.

The floods were bad. Every summer, we would have rain, of course, and sometimes there would be a weather disturbance. Every once in awhile, the water would come up to the third step on our house. My husband would fix a boardwalk, so we could get to the bridge that crossed the canal.

That was one reason the farming was so good in Ochopee. The water made the soil richer and the moisture soaked in.

Lots of times, our house was surrounded by water but it never got in our house. We got used to the water three feet deep.

Mr. Griffin's house, a little farther down, had some water in his kitchen once. The houses had all been built during high water and they knew what to expect, so the houses were well above the high-water line.

Before Barron Collier started building up Everglades City, there was no road to get there. It was like an island until Barron Collier built a road.

When my father and brother went out to look at the land at what was to become Ochopee, the Trail was very, very rough. They were still working on it, so we did see history in the making.

Barron Collier built a whole thriving town in Everglades City. We women enjoyed going over there to the Women's Club and the Auxiliary. We had lots of lady friends. Our men, Brother and Ralph, had a Lion's Club, and that met at the Rod and Gun Club.

The hotel had a famous chef from Germany. When Barron Collier was in Germany, he was impressed with this chef's cooking and brought him to Everglades City. Snooky Senghaas was his name. No one could cook like he could. We all liked to go over there for dinner. Years later, when my brother's daughter was married, they had the reception at the Rod and Gun Club. It was very wonderful.

The beautiful wedding was held at the Community Church. My sister Carrie's daughter and my granddaughter were junior bridesmaids.

As we told you, we all lived in the same house for a year. We had a big black woman as a cook. I tell you, we had a lot of good food to eat. Nancy was her name. She and her husband, Fletcher, lived in the quarters year-round. Fletcher worked on the farm. Nancy kept house for us. She did all the housework and the cooking. She made wonderful pies. She made pies of all kinds, but the best one was her grits pie. It was simply delicious; I never ate it before or since. It was like a custard pie with grits in it. She always put a stick of real butter and plenty of cream; I think she put grits in to thicken it instead of flour or starch. Her crust was as light as a feather. We enjoyed having her work for us.

We had a good store. In the store, we sold shoes, clothes, yard goods, and everything anybody would need in those days, as well as animal feed. The Indian women would often come in to buy yard goods to make their beautiful clothes. We had all we needed.

Farming isn't the easiest life on earth. Men have to get up in the middle of the night to see what they can do about a freeze or flood. Sometimes we had challenges and unpleasant trials, but we were able to meet them. Our children had the usual sicknesses. My son, Ralph, had strep throat when he was 8; we almost lost him because there were no antibiotics then.

We got our children off on the school bus every morning during the school term. My daughter, Mary Elizabeth House, says she looks back and thinks, "What a nice childhood I had in Ochopee." She lives in Tarpon Springs, now.

My son, Ralph, lives in Gainesville. He works for the Division of Plant Industry.

My daughter, JoAnn Karnes, lives in Nashville.

I think all the children look back on Ochopee as a good place to have been raised. It was a good life. We were all happy.

There were hurricanes to be afraid of, though. A bad one that I can remember was the one in 1936. Deaconess Bedell lived in Everglades City, but she always came out to our house and stayed if there was to be a hurricane, because the water always came up so high in Everglades City. She had a mission in Everglades. She was an Episcopal deaconess.

One time during this storm, I heard a BANG! I went to the back bedroom, and the wind had blown the window out. Some of the glass had blown into the toilet in the bathroom. We didn't realize it until we had trouble with that stool later.

During the lull during the eye, my husband and Vernon Vandiver went out and fixed our window and patched up Vernie's roof, as it was in pretty bad shape. We spent our time mopping floors, as the water came in under the windows and doors. It was the usual procedure to mop up constantly during a hurricane or bad storm.

I had a modern, four-bedroom, two-bath, living room, dining room, and kitchen home. Lois had a four-bedroom house with two baths, with living room, dining room, and kitchen, also. These two houses had porches on the east side and the south side. They were very comfortable homes, except for the water problem in the yard.

Now, Deaconess Bedell was my dearest friend. I'm Episcopal and so was she, so we had that in common. She came out to Everglades City in 1933. Mr. Copeland, who was in charge of everything there, was very nice to her. He told her she could have this little house to live in. Maybe she paid rent at first, but later he let her have it free. If there was anything she wanted, all she had to do was ask him for it. He knew the kind of work she did.

Deaconess Harriet Bedell had a mission in Everglades City and often visited Ochopee. She is seen here at the Post Office in 1960 with postmaster Sidney Brown and the photographer's wife *(photos courtesy Florida State Archives)*.
She also attended festivities with the Gaunt and Brown families.

She worked very hard with the Indians. She was one dedicated woman. All she got was a small salary from the Episcopal Diocese. Someone in Miami gave her a little Ford car and she had to learn to drive it. She was so proud of it. She would go to the various Indian camps to see if any of them were sick or if they needed anything. When a baby was born, she would take it a blanket and a few other little things. At Christmas time, she asked all the stores around to donate food, and she would have all the Indians together for a big Christmas dinner.

She started selling for the Indians the various items they made. She sold them for exactly what she paid the Indians; she made no profit. She went to their council meetings, and she did anything she could to help the Indians. She was in her early 60s, but had lots of energy. She had been out west to help the Indians until her allotted money ran out. She contracted TB during those years, but through prayer she was healed.

Before that, she had been to Alaska for eight years to work with the natives up there. She was a registered nurse, so she could help the natives when they needed medicine. She was a teacher, also. She told me that, when she was young, she wanted to go to China as a missionary but her mother wouldn't let her.

We used to love to hear her tell about her many experiences in Alaska. She could row a kayak as well as a native. She could drive a dog sled wherever she wanted to go. She had a little mission there, far out in the bush with the natives.

It was a good thing she came to Ochopee, because her house was blown off the blocks and all the beautiful Indian dresses, dolls, jackets, etc. were down in the mud and slime. She lost everything in that hurricane, so later on, the diocese made her go to Davenport to a retirement home.

Deaconess had dozens and dozens of Indian skirts that she had bought from the Indians to sell to tourists. They were unusual because she made the Indians stick to the old, authentic ways of making and trimming the costumes. She allowed them to use no rickrack or braid, like they use today. Back then, the skirts sold for $6 to $18. They were beautifully made, with rows of material. She could not salvage any of those Indian things, or her own belongings. She wasn't allowed to go in there for several days. By then, they were all mildewed and ruined. Her house was full of mud. Everything smelled horrible. It meant the end of Deaconess' work in our area, and the diocese sent her to a retirement home. She was a wonderful, dedicated missionary. It pained her to see the Indians in need.

She also had a little mission over in Marco for the fishermen. Deaconess went to the jail in Everglades City and held services for the prisoners. We didn't mind giving money for Deaconess; we knew where it was used.

Many a time she would stop by the house. She would be tired and need a rest. If we were eating, we would get her a plate and she'd join us.

Deaconess didn't like to cook. Maybe she would cook once or twice a week and put it in the refrigerator and eat the same thing over and over until it was gone, so she enjoyed Nancy's cooking, too.

Deaconess always came to our children's birthday parties. She loved children. She went to the school in Everglades City to speak during assemblies about her dog-sled trips in Alaska. The children were fascinated, of course.

She was a picture in her navy blue habit with a white collar and cuffs and a little white veil. We ladies couldn't understand how she was able to do things like she did, being a woman of her age. Deaconess Bedell was a very remarkable person, and we loved her.

Carrie liked to come out to visit in Ochopee. She thought it was fun out there. I guess we did have our own little world.

Well, I think that is all I can tell you. Brother, you better take over here.

James: Now Daisy, you were talking about the flood we had. Of course, hurricanes were a problem, too. The one that was the worst was Donna in 1960. She didn't damage our house, but did damage the solar water heater on the front of the house. The heater was about 16' long and the wind broke it loose so it flopped against the house with great force. This noise accompanied the hurricane wind, giving us an uneasy feeling. We had a water tank in the back of our house that was blown off its stand.

She didn't really damage any of our quarters development. She did a lot of damage at Everglades City. The water got about seven feet deep over there. It was about two feet deep in the courthouse, and that's what caused it to be moved to Naples.*

The water in Ochopee was about like any heavy rain. Since Donna came September 10, we didn't have much in the way of produce in our fields, so we didn't sustain much damage. The fields were all flooded, of course, and we had to wait for the water to recede and let the fields dry out before we could plant.

Speaking of hurricanes, in Miami we had what we call a dry hurricane. That was the 1910 hurricane that did so much damage on the West Coast and in Marco. Now, in a dry hurricane, you don't have any rain. It comes out of the south and picks up the water from the bay. That water is just salt water, and the trees on that side defoliated – orange and grapefruit trees, alike. The Peacock and Rice families in Coconut Grove, with fields of pepper plants and small vegetables on acreage, lost all their produce to the salt water.

Dick Rice had a place on the south edge of Coral Gables. There was a thick grove of pine trees there. He cleared up about ten to 12 acres and put in peppers. In the dry hurricane, he got by, because the trees gave his fields protection, as they took the brunt of the salt water.

This is an entirely different kind of hurricane. It has damaging power just like the water hurricane. When a storm came in, the Indians would get in the place of the thickest growth they could find. That is good advice for anybody that's caught in a storm with nowhere else to go.

Somebody asked an Indian one time, "How do you know there's going to be a hurricane?"

The old Indian told him, "You look at the sawgrass. When it has a heavy bloom, there's going to be a hurricane."

I don't know about that being true. Sometimes the Indians will make up anything to tell a white man!

We never had anything like that in Ochopee. In spite of the floods and hurricane threats, Ochopee was a beautiful place. It was well landscaped. We had one person who did nothing but take care of the grounds. We had our own garbage-disposal trash business. Everyone's garbage was collected daily. All the homes and buildings were well kept by our maintenance men.

Yes, we had a good thing going in Ochopee. We had good key people in all positions.

Dick Pierce was our broker. He was in charge of selling our tomatoes on the market. They were shipped by the railroad, which was not far away.

Yes, as Daisy says, we had our own wonderful little world out there. We had good homes and a good life.

Lois and I had two lovely children. My son, James, Jr., was born October 28, 1931. He is a dentist in Clewiston. Our daughter, Betty, was born November 28, 1932. She lives in Homestead and is married to Dr. Herbert Bryan. Lois always said this was like having twins. Lois had finally made a social and professional life for herself. Everglades City was alive with social affairs for us both. I was building my empire. What more could I want?

---

* *Editor's Note.* A referendum in 1959 decided to move the County Seat to East Naples but the new site was not ready until 1962.

OCHOPEE BURNS

A quirk of circumstances changed everything for us. Late one evening, a transient truck driver rented a room in our three-story boarding house.

Luckily, it was summer, and not too much activity was around. We had Mrs. McKay running the boarding house, and Orla Pettit ran the cafe. They let the truck driver have a room on the second floor.

After he went up, he was drinking and smoking in his bed. He must have passed out, because the fire in his bedroom had a good start before he ever appeared. He came running down the stairs nude, but when he saw the women, he ran back up. They were urging him not to go. A little later, he got out, but was aflame all over his body. He was jumping around like he was doing an Indian dance, because his feet got burned as he ran through the flames. One of the deputies caught hold of him, and his skin came right off in the deputy's hands. The poor fellow was so badly burned that he died next day.

When the fire started, people were screaming and running, stunned by the sight. Of course, everybody in the whole community was up and trying to help. Daisy was taking care of all the children over at her house while the rest of us were running for water.

The closest fire department was in Everglades City. There was, of course, a lapse of time before they arrived. Plenty of water was available from the canal. All the men in the settlement had been running with water buckets to and 'fro from the canal.

Finally, the Everglades fire trucks arrived. They started pumping water, but their hoses broke, so they were of no use. We tried desperately to put the fire out, but it had too good a start. One building ignited another. It was a massive fire. For a short while at the beginning, it seemed like we might put it out, but circumstances were against it.

When we saw that we couldn't save the store, Sydney Brown got all his records out of the Post Office, so they were saved.

The boarding house was just too far gone to save, so we had to let it go.

Ralph and Daisy's daughter was living in the apartment over the garage that Bill Freeland ran. All their belongings and furniture were moved out, but the garage was saved. We were thankful for that.

That store with the Post Office was a lifeline for all the people living in Ochopee. When it couldn't be saved, people began carrying out whatever they could grab and carry out. This went on for what seemed an eternity, yet it was only a few minutes.

Daisy's husband, Ralph, worked extremely hard to help save the store. We had an overhead water tank, and he used that water supply. However, it was not enough water to be of much help.

Ralph worked so hard to try to save Ochopee. He, and my sister Daisy, of course, had an interest in the operation, too, after my dad retired from farming.

The next day or so, we had to try to get organized. Ralph had to move all the business records that he had saved over to his and Daisy's house. We lost a large amount of money, because people charged at the store and all those files were burned. Ralph had a good memory, so he was able to remember what most people owed. Some people would estimate about how much they thought they owed. Some didn't know, so they didn't pay anything. Ralph was busy for days, trying to get the books in order.

The experience took its toll on all of us – especially Ralph, who had a heart attack sometime later. It was a big blow to lose those buildings – especially the inventory in the store.

The whole community was inconvenienced, because we, as well as our workers, now had to go to Everglades City to get groceries and do other trading – whether it be your shoes or a pack of needles – now that our store was gone.

Ochopee at its height had better than 200 people. We had 180 registered voters. Yes, we even had a precinct there.

Now, because of the fire, Ochopee was to dwindle down to only a few. I guess it would be a name only, if it weren't for my little shed that is the Post Office right here.

Incidentally, during the early 1950s, I was farming in Immokalee and a worker's house I had interest in burned, also.

One thing about land – it doesn't perish that easily. Besides Immokalee, I leased the Dan House Prairie and farmed it. I had also leased all of the area north of Carnestown, as well as northwest of Monroe Station, where the Seminole Park is today. Other leased farmland included some on Deep Lake Road and Bass Lakes in 1952. I had all these other interests, but my heart was in Ochopee. It really hurt when we had that fire.

Bus Station, Rooming House Burn

Ochopee Fire Causes Around $40,000 Loss

(Special to the News-Press)

OCHOPEE, May 12 — The Ochopee Trading Post, housing the postoffice and bus station, and a rooming house were burned to the ground early this morning in a fire which caused $30,000 to $40,000 damage.

A roomer in the lodging house, John Davis of Tampa, was severly burned. His screams as he fled from the room about 1:30 a.m. aroused the 10 other persons in the house and they escaped unhurt.

The cause of the fire was undetermined. It apparently started on the upper floor of the rooming house and quickly spread to the Trading Post building next door. Both were of dry pine.

The intense heat fired a garage nearby two or three times but did little damage there. Other buildings also were threatened but were saved by the volunteer fire department which came from Everglades seven miles away. Traffic along the Tamiami Trail was tied up until nearly 5 a.m.

Davis, suffering burns over 40 per cent of his body, was rushed to the Juliette Collier Hospital in Everglades and given first aid by Dr. Edward Eckel. Then he was sent to Lee Memorial Hospital in Fort Myers and put under the care of Dr. James Bradley, who later reported his condition fair. His wife came from Tampa to be with him. He was employed by the Brinson Construction Company repairing the Tamiami Trail between Carnestown and Naples.

The store and rooming house were owned jointly by J. T. Gaunt and R. H. Brown and the rooming house was operated by Mrs. Orla Albritton. She and several others in the house lost most of their belongings.

Brown said plans for rebuilding were indefinite. The loss was partly covered by insurance.

THE BIRTH OF THE SMALLEST POST OFFICE

Now we've come to the part of the story you've been waiting for – how the little Post Office came to be.

Well, you will know that it was really born out of the dying of Ochopee. As I stand here, I can see back to the buzzing daily business of the thriving Ochopee and remember the pride I felt in its growth. But quickly comes that vision of those vicious flames and all the horror of them. It was like they were laughing at me as they destroyed our beautiful dream. I remember the feeling of utter helplessness and despair when I realized our efforts were useless.

That poor man lost his life. I'm sorry about that, but thankful none of our people were hurt. That fellow never knew what he took with him. There we stood, just about all the people except children, watching the last remains become embers. The heat on our faces and the flying ashes mingled with our tears. Around us were the piles of odds and ends that were carried out of the blaze by faithful helpers. You know there was no sleep for anyone that night.

When daylight came, there lay the piles of rubble in the smoldering ashes. It all seemed like a horrible dream that would end pretty soon.

I wasn't dreaming. It was real. It was hard to believe, though. After the shock kind of wore off, I realized we still had our land and could farm when planting season came. We were lucky, really. The workers' quarters were far enough away to be safe – also, a big barn as well as our homes across the canal.

We had to begin to pick up the pieces and sort through the piles of salvaged goods. Losing that store was a big jolt, because the Post Office was in the front of the store and there was no way we could save it.

After the fire, the next day the mail came – as proof that the world goes on. We needed a Post Office. I said, "Well, we can use one of those sheds out in the field that we store irrigation pipe, hoses, and things in."

So the boys and my nephew, Ralph, went out and brought in the easiest one to move. It was mounted on a Chevy chassis, so the wheels made it easier to move. It was set down by the side of the road a little bit from where it is now. It was swept out and opened for business with mail here and there until a counter and workspace could be built.

That pipe shed worked out just fine for a Post Office, and has been here ever since that night. I understand it has been photographed many times. Some people think it was built that size on purpose, but that isn't the way it happened. Very few know or remember how it came to be.

Every time I go by it now, I am reminded of that night of the fire. My heart stops a little. Sometimes, as I look at it, knowing how famous it has become, I think Ochopee might never have gained fame as a farming settlement, but the fame came to Ochopee because of the little shed!

Some people know my name, but I'm sure not many know that it was Gaunt Company's pipe and hose shed that became the Post Office after the fire. There were several sheds in our fields, but that particular one happened to be the one chosen.

Life is full of strange circumstances; don't you agree?

THE PAINFUL END

We did close down our packinghouse and living quarters for our workers at our Ochopee settlement after that fire. But this fire didn't stop us from farming that season, though we felt quite down about our great loss.

The season following the fire, I contracted some of the crew leaders in Immokalee. They would gather up their 25 or 30 people and bus them over to our big fields in Ochopee. We carried on our operation that way. Lois and I still kept my home in Ochopee, and so did Ralph and Daisy.

I had a packinghouse in Immokalee at the same time as Ochopee, so we used that after the fire.

I had started farming under the name of J. T. Gaunt Company in Immokalee in 1939. My partner up there was C.J. Jones, who had the big lumber mill. Our company was named Immokalee Growers.

Roy Miller was a partner, also. We leased farms all around Immokalee and other places. Dad always had more than one farm going. I guess that's called not putting all your eggs in one basket. He taught me well.

After the year of the fires, we had some other bad luck in Ochopee. There were some bad weather conditions, low market prices, blight, and drought. Sometimes it would rain too much, with no sun, and the blight would come on. The blight made the tomatoes look watery and gray. It seemed we couldn't find a spray to stop it. We lost fields of tomatoes. If we shipped them, they would go into the market rotten.

It took so much money to lay out a farm as big as this operation. It took a lot of money to get the land prepared; get the fertilizer, the seed, the insecticides; pay labor, the upkeep of all the buildings, and the operation of the packinghouse. What really happened was that the tomato market priced itself out of business.

Before World War II, labor was $.75 to $1 a day. The cost of labor went up when the war came on. The high cost of fertilizer for that sandy land, added to the cost of spray for the blight, added to the rising labor cost, caused the Gaunt Company in Ochopee to begin to draw operations to a close after the drought which followed the fire.

The next five years of the Gaunt Company were not good ones. After the 1953 fire, Daisy's husband, Ralph, had a heart attack and died in 1955. Dad died in July 1960. Hurricane Donna came September 10, 1960. Mother died in 1970.

Farming is a gamble. Things have to work out together for money to be made. When there is dry weather, too much rain, or a freeze, a farmer can't do anything about natural causes for crop failure. He just has to try again.

"You either have a lot of money or you don't have any!" is what Daisy always said about farming.

By 1965, Gaunt Company closed up the Ochopee operations completely, but I lived there awhile. We ended up selling some of our land. Part of it became the Bass Lake Estates. We still had our homes on sizeable plots for awhile.

Soon, the Everglades National Park was buying up the land. Today, Daisy's house is still here, over across the canal, and a few of the others, but they belong to the park.

Ochopee isn't much of anything, now.

In 1962, I moved my family to LaBelle, since I was farming with a man named John Campbell from the East Coast, as well as farming between Immokalee and LaBelle. With farm operations so widespread, I was driving in all about 200 miles a day, and it got to be a chore. It was greatly improved when I moved my family to LaBelle. Daisy later moved to LaBelle, also.

In 1972, Mr. Campbell and I retired from farming. I was 72 years old, and I had farmed all my life and loved it, in spite of the setbacks that came. I did well. After my retirement, I kept active with my own little private garden. I also participated as a trustee in the Lee County Electrical Cooperative, which had been in existence for 34 years.

I was president of the Cooperative Board of Trustees from 1956 to 1981. Back in 1975, I was the representative for Ochopee and Everglades City. From the beginning of rural power, I was sure rural electrification was good for progress in business, farming, and residential area. I think I did my best to improve the community around me, just as Dad always did.

Well, I guess that is just about all of the story of how our little shed got to be the smallest Post Office, here in the Everglades, in a place called Ochopee.

MEMORIES OF OCHOPEE

James T. Gaunt, Jr., D.D.S (son of James Gaunt, Sr.)
Betty Gaunt Bryan (daughter of James Gaunt, Sr.)
Herb Bryan, Ph.D. (son-in-law of James Gaunt, Sr.)

James, Jr.: It's no wonder the black people who lived in Ochopee thought Daddy was so wonderful. Our company provided power for everybody out there. They had landscapers who took care of every yard and all the land in the town of Ochopee. The company built their houses and maintained them. That applies to the white workers as well, except for the Cail house. They built their own home.

There was free garbage collection for the community. Everyone could charge their groceries at the store. When the store burned in 1953, some didn't pay my father what they had owed for years. The black people who worked for my father stayed the year-round. His settlements were more like a plantation. My father provided for all of their needs. When they got sick or cut up or knocked each other around, he would take them to Everglades to the doctor. They just grew to expect that.

Betty: When any of them got put in jail for fighting or whatever, Daddy always sent somebody on Monday morning to get them out, and they would go to work.

James, Jr.: Ochopee was a very paternalistic society. It was a very special, rather isolated type of world.

Betty: Our Daddy provided many jobs to drifters during the depression years as well. Very, very few people – except a pioneer like my father – would have taken the risk during those years to move out to an isolated, God-forsaken area like the Everglades used to be. Keep in mind that this was before insecticides to combat the mosquitoes. It wasn't very easy living out there in the early 1930s.

James, Jr.: In Ochopee, there were about 20 kids around the same age. We all played together. On weekends, our family used to go to Marco Beach to swim. When we were through swimming, there wasn't any place to change from our bathing suits and we sometimes had to come home wet. If it was summer and the wind was blowing from the land, the mosquitoes and sand flies were so bad we had to run to the car as soon as we got out of the water. We often changed in the car. That was inconvenient, so we always asked our father to buy us a home in Naples on the beach. We felt like that would be ideal. We'd have a place to spend the weekend – to swim and have fun.

Maria Stone ~ OCHOPEE: The Story of the Smallest Post Office

After a certain event, we found out why he never would do what we wanted him to do. Our mother was a good friend of Mrs. Briggs of Briggs and Stratton of Naples. They had a nice home on the Gulf. One day they had called and we went over to their place. This was in 1936, right after there had been a bad storm. The storm had carried sand all through their house during a tidal surge. The first floor was destroyed. There was sand three or four feet deep inside their house, and the furniture was all smashed. Mother played the piano so well, and so she was upset over their piano being torn up. So my father told us right then while we were viewing this destruction, "This is why we will never have a house on the water. When a bad storm comes, a tidal wave will surge through the house and destroy it, just like this house. The same thing happened to your grandfather's house in Cutler, south of Miami." Water went over Grandfather's house during a storm. It didn't tear up the house, but it smashed all the furniture in it.

Our father always remembered what had happened to his family home, so he would never have a house anywhere near the water. That day at the Briggs' home in Naples, we understood his reasoning. Our father wanted to live inland. Ochopee was far enough inland, but during Donna, the houses in nearby Everglades were torn off their pilings because of the water surge that went through there.

From the time that I was 10 years old, I worked at the packinghouse. That was during the depression, and I remember people being out of work and times being bad for a lot of people. We had quite a few seasonal workers in the fall and spring. They came there for a job, but could have done much better elsewhere if it had not been such hard times.

My cousin, Sid Brown, and I used to work on Saturdays. We made boxes up on the second floor of the packinghouse. We had a big machine that made lugs in which we packed tomatoes. We put the boards in, stamped down on it, put the nails in, turned it on the side, and put on the ends. After that, the lugs were ready to go to the pickers in the fields. I worked there for $.50 an hour. I should not have worked there because many people needed a job, but I had the job just because my father owned the operation.

The packinghouse foreman was Matt Bonds. He was also foreman in Immokalee when the operation was moved there. In Ochopee, we had one man who took care of the mule barn, which was a very large building. He took care of all the mules. He had come down from Georgia. We found out that, when he was younger, he had killed a man. To get away from going to jail, he came to live with us in Ochopee. He had an apartment in the barn. Mr. Pat is all the name we ever knew him by, but he was related in some way to the Brown family. He lived in the barn until he died.

Mr. Pat had a bad leg, and he dragged it as he walked. From anywhere in the packinghouse, we could hear him walking. When he walked across the upstairs floor, you could really hear him drag that leg. In the packinghouse, he did things like pasting the labels on the lugs. He also put the lugs on the truck, and they were then carried to the railroad side in Carnestown and shipped to New York.

The black settlement was not far from the mule barn. The black people were frightened to death of Mr. Pat because they thought he had killed a black man in Georgia. The black people wouldn't go near the barn when he was there. They wouldn't touch anything in the barn either, because they said, "Mr. Pat will git me." After Mr. Pat died, they would say they could still hear him walking in the packinghouse at night. They wouldn't go near it. "Mr. Pat's ghost is there," they would say.

I remember one time in the packinghouse Mr. Pat stepped in a bucket of paint and turned it over. He got real mad and cussed. Everybody in the packinghouse said, "Mr. Pat is bad. You better not go near him. He'll git ya."

Betty: Most people who settled in the areas of Chokoloskee, Everglades, Ochopee, and the whole area were people who were really nonconformists. They wanted to live their own lives. They didn't want anyone to tell them what to do. They wanted to create their own world.

Daddy was that way, to a degree. He was very innovative. He always wanted to explore and try new ideas. He had a pioneer spirit. That is why he wanted to come to Ochopee. He was looking for a new challenge. Most people who were well established and had money would never have come to Ochopee. They would have stayed where they were. Our father wasn't like that. Most people who came to Ochopee to work needed a hand to lift them up and our father did that for many people.

Daddy was always looking for new ideas, and between the fall and spring crops in 1937, Daddy took his two foremen with him and went on a trip out west.

James, Jr.: When Barron Collier built Everglades City, he brought a chef from Germany who had been a chef for the Kaiser. He was Snooky Senghaas, who became the chef at the Rod and Gun Club in Everglades City. Also, Mr. Collier brought Mr. Hans Schmeisser as the golf-course designer and horticulturist. After Hans quit working for Mr. Collier, he became a foreman and farmed for Daddy in Immokalee. Daddy had a Japanese foreman in the 1930s in Ochopee. His name was "Boots" Si Yon. In 1938, Boots' wife and son went back to Japan and were caught there when the World War II started. They didn't get to return to the United States until the war was over. After Ochopee, Boots Si Yon went to southeast Georgia and grew lettuce.

But, getting back to Daddy's trip: Boots, Mr. Schmeisser, and his son, went with Daddy to Colorado and Washington State. While out there, they visited all the national parks. When Mr. Schmeisser left Immokalee, he looked after the golf course at the Hollywood Beach Hotel.

Betty: I think my Daddy knew everyone of importance in the State from the early years up through his last years. It was remarkable, but he never forgot anyone's name. He had a fantastic memory. He was gentle, generous, and honest.

Herb: He had a great deal of vision about development and progress. He realized the need for electricity. He pushed to get electricity in the whole area of southwest Florida. He was one of the prime forces working to get the Agricultural Research Center in Immokalee. Paul Everett was director of that center for many years.

Betty: Daddy was also very environmentally minded, which one wouldn't expect, because most farmers get a "black eye" with the environment people. I remember when the engineers first started dredging all those canals in Everglades. Daddy said, "This will ruin the whole area," and it seems that it did just so. Various farmers get the blame, but the blame belongs at the door of the politicians who directed the Corps of Engineers to drain the Everglades for the use of canals and dikes. It may have benefited the future developers.

Herb: The real reason for the draining was for flood control. When all those people drowned in Okeechobee during that storm, there was need for flood control. We realize now that it ruined a lot of land and this river that comes down the center of the State into Lake Okeechobee as well as the areas to the south. The other thing that Dad had a great objection to was the cross-Florida barge canal. He felt that would ruin the underground river that brings water into southwest Florida. He understood the geology of southwest Florida.

Betty: When my brother entered ninth grade, he moved to Miami to stay with relatives and go to school. The next year, my mother and I moved to Miami so that I could go to school. Daddy stayed in Ochopee. At that time, he also had farms in Immokalee and elsewhere. He would come to Miami every weekend. Therefore, our knowledge of Ochopee faded after 1946 because my brother and I never returned to Ochopee except for a day or two during our holidays. We had entered college after our high-school years in Miami.

James, Jr.: When our father started farming in Immokalee, it was an old, wild west town with cattle drives and jukes and dirt streets. My father had to buy two blocks of land, divide them up, and build houses for his black people who were to work his fields. This was the beginning of the change in Immokalee, and also the change of the County's focus from citrus/cattle to large-scale farming.

Betty & Jimmy Gaunt as children in 1940 and as adults in 1952.

MEMORIES OF OUR FATHER IN OCHOPEE AND THE SMALLEST POST OFFICE

Dr. Herb Bryan and Mrs. Betty Gaunt Bryan

Herb: Mr. Gaunt was a genuine farmer. I remember that, when the Tamiami Trail was built, it was raised above the normal land level. That was before it was even paved. What I'm going to relate to you happened in Ochopee as early as 1929 and on through the very early 1930s – and maybe it happened more often than those years I've mentioned.

There was very little traffic on the Tamiami Trail those days, so Mr. Gaunt plowed the shoulders of the road and planted two rows of tomatoes on each side of the road for perhaps two to four miles. The reason for doing this had to do with the water level during the rainy season. The water got high sometimes, and the tomato crop was lost in low areas of the fields because the soil level was very low and close to the water table. Up on the roadside, being above the water table, the crop would survive; therefore, if there was an usually wet season, the roadside crop would be worth a lot of money. If it was four rows – two rows on each side of the road – that would be approximately three acres or more of tomatoes per mile, assuming from the general yield during that time that the amount of tomatoes realized would be perhaps 4,500 cartons per mile. That would translate at $10 a carton into $45,000.

Betty: Daddy said those were probably the longest tomato rows in the United States – and probably the world.

Herb: This quote – about the mile-long tomato rows being the longest in the State if not the World – appeared in the Florida State Horticultural Society Magazine. The statement was made by one of the past presidents at an annual meeting of the Society. He was the Director of Research for the University of Florida Experimental Station and the Institute of Food Science and Agriculture.

Mr. Gaunt used to visit us quite often in Homestead. He liked to go out with me to the plots where I was doing research, or to grower fields, and also to the Research Center with me. He was always interested in new developments in farming. In 1987, at Thanksgiving time, about three weeks before he died, he was with me at the Research Center to observe a new farming technique. A planting machine usually does three rows at a time. The burners on the machines burn a hole in previously laid plastic mulch. Next, a seed is deposited in the hole, along with peat and vermiculite, which is a clay material that acts as an insulation. Our next procedure involved spraying liquid plastic in the holes. By using the plastic spray, a grower would only have to irrigate every three or four days. Without the plastic spray, the wind quickly dries out the planting, and a grower is required to irrigate the tomato field every day for one-half hour until plants come up a week or ten days later – depending on the season. Naturally, our procedure decreases the use of water required by the farmer. This research produced a more uniform stand of plants and better yield of fruit for the farmer as well. This is one of the many projects that we were trying to do on a grower's scale for field use.

One day during Mr. Gaunt's visit, he was at those research plots with me. I was taking pictures of the work. As Mr. Gaunt walked along, he noticed that the hose of the pressure sprayer we were using was dragging over the plant beds and catching small rocks which tore holes in some of the plastic mulch. Mr. Gaunt came along in back of the sprayer, picked up the hose, and carried it along so it wouldn't damage the plastic mulch and disturb the seed. We didn't ask him to do it; he just had a sixth sense about farming. He was a natural-born tender of plants. He was always keeping up on new developments in farming procedures, so he enjoyed coming to the Research Center.

He was also interested in our project of seed germination. For that research, in special plots before we planted the seed in the plastic mulch, we used a machine to distribute jell which coated the seeds. This research proved that jell protected the seeds and held moisture as well. Dad Gaunt was well aware of the value of this germination procedure to increase the yield of a crop.

In the 1940s, Mr. Gaunt had problems in his Ochopee tomato fields with late blight. The chemicals that tomato growers have available today prevent late blight from being a serious problem. It has been in use for the last 15 years. Mr. Gaunt often lamented to me that he had only nicotine sulfate and copper to combat blight, which wasn't adequate to control the blight in his Ochopee fields.

In my early years, I farmed with Mr. Gaunt in Immokalee. This was in the fall of 1956 when I got out of the Service and college. We had about 75 acres of watermelons that were a very successful crop. Then, in the spring, we had 30 acres of cucumbers and 30 acres of watermelons. About a week before harvest, it rained 17" a week for three weeks. The fruit floated down between the rows. We were unable to harvest anything. I decided right then that that was a message for me to go to graduate school. I remember one day while I was working on some old farm equipment and a truck during that rain, I was all greasy and dirty, of course. When I got out from under the truck, Mr. Bernie Serkin* was standing there talking to Dad Gaunt. Mr. Serkin financed some of the farming operations in Immokalee.

Dad introduced me to Mr. Serkin,* and he said, "Young man, I'd like to buy you a suit. I know that you young people have hard times out here, and I want to help you out." I was insulted, probably because I had just finished that frustrating job of trying to keep our equipment working on top of our crop failure. I didn't feel like his description fit me. That's when I went away to Cornell University. I majored in vegetable crops and received a Masters and Ph.D. degrees in horticulture. Since then, I have done research to improve the farmers' methods of growing vegetables. To be a farmer, one has to have faith and stay in it long enough to let the good years offset the bad years. After that rain in 1956, I didn't choose to wait for the good years.

Betty: My father never, never seemed old because he kept abreast of current developments in all subjects – but especially farming. He never, ever mentioned that he didn't feel well. He was forever eager to read the daily newspaper and periodicals. He kept up on events in the world, also. In his farm villages, he never took advantage of black people, Indians, or any other people he dealt with. He was very much opposed to others who did it. The whole community of Ochopee loved my father. If ever they needed a helping hand, he extended his. I can't say that about everyone who lived in Ochopee. Not just because he was my father, but he was the most gentle, generous, honest, kind, sincere man you would ever meet. He dealt with all people fairly. People who knew him will tell you that. He was a real gentleman.

---

* Mr. Bernie Serkin owned the Ochopee Post Office and the land surrounding it.

FRIENDS & NEIGHBORS

Chief O. B. Osceola: MR. GAUNT'S FARM

My family worked on the Gaunt tomato farm when I was a small boy about 8. My brother and I would run away from the fields because we were tired. By the time we walked back to our camp, which was a mile away, our father (Corey Osceola) was waiting there and he would take us back to the fields to work.

One time, my job was water boy. I had to walk a long way back to the mule barn to get a bucket of water from the well there. It was heavy, but I carried and dragged it back to my family working in the tomato fields. By the time I'd get to them, the bucket would be about empty. My brother and I would get in tomato fights when my dad wasn't around. When I was small, one of the men let me drive a mule behind a fertilizer spreader. The mule didn't understand my signals. By accident, I pushed the lever that lowered the machine and some of the plants were torn up.

One thing I remember that looked so funny about those mules: Mr. Gaunt had Ed Frank of Naples make special shoes for the mules' feet so their feet didn't sink into the mud in the fields.

Ochopee was a good town. It was a big town, back then. We kids liked to buy bologna and cheese in the Gaunts' big store that Mr. Brown ran. All those people in Ochopee were good people.

**Local Indians helped to pick tomatoes for the Gaunt Company.
This photo is in the Bedell Collection at the Florida State Archives
and the handwriting is that of the Deaconess.**

Charles Price, Jr.: MY FIRST JOB

My great grandfather was Walter Haldeman. My mother was Bruce Haldeman's daughter, Florence Haldeman Price. My father was Charles B. Price, Sr. All my life I spent winter vacations at my grandfather's house, which is still by the pier. They owned the whole block. Walter Haldeman came to Naples in 1885, and, for many years, I came to Naples with my mother, my father, and my two sisters, Anna Haldeman Price and Violet Haldeman Price.

In December 1945, when I was on terminal leave as a lieutenant colonel in the First Armored Division in North Africa and in Italy, I came to Naples. I thought I might like to live here from December to May. I wanted to earn enough money to go to Jamaica. I tried to get a job seining fish, but the weather was poor and the fishermen, George and Rob Storter, didn't need me.

I heard of this large tomato operation in Ochopee owned by Mr. James Gaunt. He had about 2,000 acres or more in tomatoes. There was a large packing shed as well as a regular village in Ochopee. I drove out there and asked for a job. I was hired as "low man on the totem pole." The tomatoes were picked "dead green". I tried to get a job picking them, but was told that was "stoop labor." They wouldn't hire anyone for that except Blacks, Mexicans, and a few nearby Indians to do the actual picking. The people working on the conveyer-belt line grading and packing tomatoes were all Whites.

My job was at the beginning of the tomato conveyor belt. I dumped the dead-green tomatoes on the conveyor belt. They had been picked and put in 60-pound crates, which they called field crates. The tomatoes came off the belt, and my job was to pick out any tomatoes that were slightly pink or any that had blight or other defects. The pinks went to Miami and were sold as ripe tomatoes. The culls were thrown away.

As the tomatoes went down the packing line, they were immersed in a solution of hot wax. Those tomatoes were graded by size and wrapped in tissue paper. After that, they were packed by hand in 20 or 30 pound lugs. They were called five-by-fours, six-by-sixes, seven-by-eights, or however many rows they would pack in those little boxes. The little boxes were put in cartons, labeled, and loaded into a boxcar to be shipped up north. Obviously, leaving as dead-green tomatoes, they ripened as they traveled to northern cities.

I was quite interested in the tomato business. I thought of the possibility of going into it myself, but I was told that it was a very expensive operation. It was luck, because the next year the Gaunt operation had a blight and just about all the crop was wiped out so there was no profit.

For the time I was working in Ochopee, I had a car to drive until my parents and grandparents went back to Kentucky. After they left, I had to hitch a ride to Ochopee every morning. It was close to 40 miles. Gas rationing was still in effect, so there were very few cars on the Tamiami Trail. I had to be at the packing plant by eight o'clock.

The pickers always had to wait until the dew dried up before beginning picking. I seemed to be lucky, as I was able to arrive there without much trouble by eight o'clock packing time. I rode with all kinds of people. One day after I had been working there two or three weeks, I got out on the road and there was no traffic at all. I started walking and finally a little coupe came along with two Indians in it. They didn't say anything, but opened the door, and I got in. I noticed that there was no floorboard and the crankshaft was going around under my feet. I didn't dare put my feet down, so I held them up the whole ride. They drove up to Weaver's Camp, I believe it was called. They stopped and let me out, then they took off across the Everglades, bumping along as they went. I couldn't believe what I saw.

They left me on the Tamiami Trail alone, and the prospect for another ride seemed hopeless, so I went in and bought a ticket in Weaver's Station to ride the Trailways bus. The reason I didn't ride this bus every day was obvious. I was getting $.85 an hour and the bus ride cost $2. It also cost $.85 for breakfast at one of the little restaurants in Ochopee, so I ate breakfast at home and packed my lunch, since there wasn't any cheap place to eat out there. Some people bought something to eat at Gaunt's big, general store, which had a Post Office inside at that time.

I guess I worked there six weeks or a little longer. One day as I pushed a cart of tomatoes toward the belt, they dumped right at the feet of a tall man. It was Mr. Gaunt, the owner of the operation, who was there to check the packing process that day. He said, "Son, I don't think you are quite cut out for this job." I quit that day. That ended my tomato business, and I went on to Jamaica.

Col. Frank Tenney: THANK GOODNESS FOR THE LITTLE OCHOPEE POST OFFICE

We came to Naples in 1961, a year after Hurricane Donna. We had been down in the Florida Keys in a fairly new Oldsmobile car. My mother was with us, and we had gone down to the Keys for a three-week vacation. We had our car loaded because I was headed for Colorado Springs, Colorado. I was in the military in the capacity of a colonel in the Air Force at the time. I had come back from Hawaii and was on a three-week vacation in the Florida Keys.

Since I was to retire rather soon, I had to look for a place that we might want to live. I had owned some lots in the Keys for sometime. While on vacation there, I saw what Donna Hurricane did to the Florida Keys, and I said, "There is no way I would build down there on the land we owned."

We left the Keys just before the end of my leave, which was just before Thanksgiving. Before leaving the Keys, I had tried to ship my diving gear and such from Marathon. All this equipment was in a couple of 60-pound wooden boxes. I supposed I could mail them at any Post Office, but when I went into the Marathon Post Office, they said, "We can't mail 60-pound wooden boxes. You will have to go to a Class C Post Office."

Well, I loaded them back in the car and it was so loaded that it was practically down to the springs and nearly dragging. Nevertheless, we started out at daybreak and came up north on the Tamiami Trail through Ochopee about nine o'clock that morning. Suddenly, I saw this old man* out putting up the flag in front of a Post Office. There was the little Ochopee Post Office. I remarked, "Oh, look! If there is ever a Class C Post Office, that has to be it." I drove in and asked the old gentleman if he could send a couple of 60-pound wooden boxes for me from Ochopee to my home. He said, "I'd be happy to do it."

I was so glad to see that little Post Office. Those boxes got home before we did because we came on into Naples. We liked it so well that we bought a lot to build on later. The rest of the story is history. That's my memory of the first time I saw the Ochopee Post Office.

---

* The old man at the post office was Mr. Sidney Brown, the postmaster.

Bruce Warren: THE MAIL ROUTE

My father, "Bony" Cecil R. Warren, came to Miami sometime between 1923 and 1925. He was a reporter for the *Miami Daily* newspaper for over 20 years. He was also a radio commentator for WIOD.

My brother, C. Rhea Warren, a herpetologist, and I used to collect snakes in the Ochopee area. My brother was the first scientist ever to have four reptiles/amphibians, new to science, named after him.

On an assignment to Okeechobee in 1928, after the storm drowned 1,500 people, my father was there with President Herbert Hoover. During conversation, President Hoover patted my father on the back, saying, "You are sure bony, aren't you?" For the next 20 years thereafter, my father, as a reporter, went by the name of "Bony" Cecil Warren. That nickname was given him by President Hoover.

My father was a campaign manager for U.S. Senator Spessard Holland. Also, my father was campaign manager twice for the Governor of Florida.

As I grew up, I didn't want the pressure of conforming to society like my dad or my brother. I liked the freedom of the Everglades where we often went as young kids to catch snakes to sell in Miami.

After high school, I left home for a carefree life in the woods near Monroe Station on Loop Road. While I was there, I noticed that there was no mail delivery up and down the Tamiami Trail. Whoever could, if they could get to the little Ochopee Post Office, would pick up the mail and bring it back for everybody along the Trail. Usually, it was brought to Monroe Station and left by the register until someone who had mail happened to come in.

Now Old Benny and a couple of other retired people out there were living off their Social Security checks. These people weren't getting their checks on time. Perhaps they were lost or they would be delayed in the handling of the mail.

I made up my mind to do something for them. One night, I went up to the Miccosukee Indian Village and used their phone to call one of my father's friends. To the operator, I said, "My name is Bruce Warren, and I want to call Senator Spessard Holland at the Senate Office Building, Washington, D.C." To my surprise, he picked up the phone. I said, "Senator, you don't know me from Adam, but my father was your campaign manager." My father was deceased at that time (1960), but the Senator asked, "How is your mother?" After a short conversation, I said, "Senator, I want to do something about trying to get mail delivery for these people who live up and down this Tamiami Trail." He said, "You write the particulars to me. I'll see what I can do."

Now these people around Monroe Station where I had been living didn't know I had any connection. To them, I was just a swamp bum who had run away from home at 16. The people on the Trail didn't know about my family background. I was just little old Bruce, the snake-hunter kid who lived in a chickee and drank too much beer sometimes.

The next day, I wrote the senator. We exchanged letters several times, then one day a man drove down Loop Road. He was in a new car hunting me. He said, "I'm from the Post Office Department in Tampa." I took him all along the Trail and Loop Road and showed him all the junk trailers and chickees and the run-down shacks of everybody that lived along the Trail. "Okay," he said as he drove off. "We'll see what we can do."

About three months later, a mail route was started up and down the Tamiami Trail from the little Ochopee Post Office. Before that, everybody up and down the Trail had been driving to Ochopee Post Office to get their mail, which was about 50 miles away, in some cases. Either that, or, as I said, wait until it was brought by somebody coming east on the Trail from the Ochopee Post Office.

Now, this was in the 1960s that the people along the Tamiami Trail began to get daily mail delivery. The Ochopee Post Office for that area had been established in the James Gaunt settlement in about 1928, so for 30 years or more the people on the Tamiami Trail had had no mail delivery. Monroe, Dade, and Collier Counties joined together at one point, and the Ochopee Post Office services more than 200 citizens in those three county areas.

During that time, I got the *National Geographic*, also, to come out and do a feature on the Everglades and the people there. They wanted to give me some publicity, but I refused it and asked them to feature my friend, "Gator Bill," instead. These projects that I'm telling you about drew attention to the Everglades and the people along the Tamiami Trail in the 1960s. I did all this unselfishly because I think the greatest thing is to do something and not let them know where it comes from.

Mr. Meece Ellis: REMEMBERING MR. GAUNT

Mr. Gaunt's father was called "Ol' Blessed" by the black field workers.

I remember that bootleggers of that time would buy a load of tomatoes from the Gaunt packinghouse. The tomatoes were then put in the front of the boxcar to conceal the load of liquor that they were shipping out to the north. The load of liquor was hard to locate, as the Gaunts were shipping thousands of carloads of tomatoes out of Ochopee during those years. The revenuers finally found out about the bootleggers' trick of hiding liquor behind tomato cartons.

The Gaunts were never aware of this being done.

Lee Hancock: GAUNT'S MANY FIELDS

I remember back of my daddy's house (Ken Hancock in East Naples). Mr. Gaunt had leased some very large fields. He built a packinghouse back there. I sometimes worked for him.

Author's note: It has been established that Mr. Gaunt had vegetable fields all around Collier County as well as Dade and Hendry Counties.

Jerri Fish: I COVER THE EVERGLADES

When Sidney Brown was postmaster of the Ochopee Post Office, Mr. Luther Bates had the contract, right from the beginning, of the mail route. He and his wife drove it together. She helped him because he was getting up in years and had a heart problem.

He hired me to drive for two years, as they were not able to do it, and then I took the mail route. I have driven this route for 22 years, six days a week. I have a smashed front end on my truck from hitting a deer. That's my fifth deer. I can't prevent hitting them. The police say, "Jerri, you are a better deer hunter than most who try to hunt deer." That's what they tell me when I make out a report.

Every day the mail comes by truck from Ft. Myers at 7:45 a.m. to the Ochopee Post Office. It is in big bags, and the postmaster sorts it and readies it for my route. My route goes like this: From the Ochopee Post Office, I come 41 to Highway 9. I go 29 to Copeland and do a pick-up, then go on to Jerome, backtrack, then go to Birdon, then on to Wagon Wheel Road, and back to 41. Down 41 to Turner River Road; I go up all the Turner River Road, backtrack to 41 again, then go to Burns Road, then go back to 41 all the way to Shark Valley and Dade County delivering mail all along the Trail. I turn around and go back 41 and do the entire Miccosukee Indian Reservation. I come back out on 41 to the Tamiami Ranger Station, then down through Loop Road to the Environmental Education Center. Then backtrack to 41 again, and come on to the Ochopee Post Office with a load of out-going mail. The big truck comes back for out-going mail at about 4:30 or 5:00 o'clock.

In the 22 years, I've seen a lot of things that would frighten most people. On Burns Road, a girl's body was hid behind a log not far from the road. One day driving through there on my route, I saw all these police cars at a little curve. I hadn't seen a thing, and I had driven through there every day. I never even thought about anything like that on my roads. The guy who did it confessed, so he wasn't running loose. Every time I go around that little curve today, I think about this. I had been passing her body and didn't even notice her behind that log.

About ten years ago on Tamiami Trail in Collier County, someone had killed a man and had thrown him out of a car. A trucker had put a coat over him. I had to drive around that dead body and continue delivering mail at my boxes. They caught those guys that were responsible, too. Nineteen years ago, after I started, body parts were found in a garbage bag on Loop Road. I happened along on my route just about then.

The Everglades seems to be the dumping ground for bodies because it is so remote.

The settlement of Pine Crest had Gator Hook Bar.

At first, I delivered to 25 or 30 families in Pine Crest, but that town died out when the Park took over all the land around there. Pine Crest was a place that governed itself. I only went through there to deliver mail. At first, I was afraid, as I was just a young girl back then and a little nervous.

I've never carried a gun but my brother worried about me going alone on those roads and in the woods. As I say, at first, sometimes I was a little leery but now I have no fear. People on my route are some of the best people I've ever known. If I'm in trouble with the truck and miles from nowhere, anyone on that mail route will help me.

I've been in some bad storms. I drove when water was so high it would come in the doors of my truck. I'd wonder if I could get in and out of the back roads. I always did. In bad rain, I'd sometimes pull over by the roadside and hope it would pass.

In 22 years, I've never failed to go. I don't think I'd do anything else. I'm used to being on the road. I never have a vacation. There's nobody to do it. My route is one nobody wants to take on. No woman, especially, and not even any man. I cover the Everglades. My route covers the outlying areas, remote as they are.

Maria Stone ~ OCHOPEE: The Story of the Smallest Post Office

I came here from Kentucky 24 years ago, and lived in Naples. I drive the 40 miles out to Ochopee from Naples, do the route, and drive 40 miles home to Naples. Ten years ago, I moved to Copeland. It is much easier to get to work.

Now the Everglades Post Office serves Everglades. Copeland Post Office serves Copeland. But Ochopee serves the whole Everglades area. I do 160 boxes. We are putting in 36 more boxes; that will make me 196 boxes to serve.

There is no way to get rid of the Post Office in Ochopee. It is a landmark. If for no other reason, it should always be kept for the history behind it. The people it serves love it. It is a special little thing. Having the smallest Post Office in the United States distinguishes us from every other place.

On my route, I've never seen a panther for sure, but I've seen about every animal and bird native to the Everglades. I see snakes all the time.

All maps have points of interest marked on them. It seems to me our Florida map should have the little Post Office marked as a place of interest.

The Everglades seems to be the same, but very few old-timers are left on Loop Road. One is Old MacDonald and his dog. He's lived in a converted school bus. I brought his tobacco and a loaf of bread when he needed it. A man named "Skip" who lived on Loop Road wrote songs about the Ochopee Post Office and my mail route. He played the guitar and sang them in Gator Hook Lounge.

Another interesting man I met at the Ochopee Post Office. The three acres of land are owned by Mr. Bernie Serkin. The Postal Service leases the building. He was as proud as anything of that little Post Office. He was a friend of Evelyn Shealy, the postmaster, and he visited her often at the little Post Office. I got acquainted with her when I began the route. Evelyn was a good, good person and did a lot for the entire community. I loved her. She was not only someone I worked with but a dear friend. Everyone who met her loved her.

One man on my route that I will never forget was Irvin Rouse. He wrote the famous song, "Orange Blossom Special," so you see, I've met a lot of interesting people in my years on the route.

I believe I do a good job. I'd go out of my way to help anyone on my route. I do almost everything out of my truck that can be done in any Post Office. I can sell money orders, stamps, and so forth. The people are so far out I've got to do these things for them.

The little Post Office is unique. If we postal people had to move to a cold, concrete box, it would be sad. I have an attachment to the little Post Office. I think everyone does who goes there, but first of all we must remember that it is a working Post Office. It is not a tourist attraction. It serves people in a large area, but it is a charming place to stop beside the road to Miami.

Author's Note: Jerri Fish drives 123 miles a day, six days a week, excluding holidays, for 52 weeks, and for 22 years this makes a total of approximately 844,000 miles she has driven her route.

Maria Stone ~ OCHOPEE: The Story of the Smallest Post Office

Evelyn Shealy was Postmaster from 1971 until she died in 1991.
She is seen above at the counter inside the front door.

SHEALY FAMILY & NEIGHBORS

MEMORIAL TO EVELYN SHEALY

This section of the book is dedicated to the loving memory of Evelyn Shealy, who was postmaster from October 1, 1970, until her death in July 1991.

Memorial to Evelyn Shealy
By Maria Stone

When you stepped up to the Ochopee P.O.
You heard Evelyn Shealy's cheery "Hello!"
She greeted each stranger like a special friend
Then sold them stamps or post cards they wished to send.

Evelyn loved her job and she let it show –
Twenty years shut inside the tiny P.O.
There were days of driving rain and scorching heat –
Evelyn never changed; she was always sweet.
Often sitting alone in that sawgrass world –
Unafraid of snakes in the letter box curled.

She gave help to those in need both far and near.
To all who knew Evelyn, "She was a dear."
This postmaster girl in her little P.O.
Sent out to all of the world a special glow.

Natives and travelers came oft by the score
And never forgot the smiles Evelyn wore.
Took pictures and her blessing and went away
All charmed and happy by what they saw that day.
This smallest Post Office in the U.S.A.—
In all the world, nothing like it, people say.
Their fame spread far beyond the River of Grass
On Tamiami Trail where the cars whiz past.

'Twas July '91 Evelyn did pass.
Left her little P.O. alone in the grass.
A goodwill ambassador and loyal friend,
Evelyn gave of her best right to the end.

She is gone but in our hearts she will remain,
And the little Post Office echoes her name.

Evelyn Shealy:
MY YEARS AT THE LITTLE OCHOPEE POST OFFICE

In 1990, Evelyn got an award for the Outstanding Postmaster of Florida, according to her sister. Ochopee has changed so much through the years.

Evelyn: Loop Road (private) is located about 35 miles from the Ochopee Post Office. It is State Road 92, I believe. It runs off the Tamiami Trail at 40-Mile Bend. There is an Indian church there. The three counties of Dade, Monroe, and Collier join there. We have served the three-county area for more than the 20 years that I know about.

On October 1, 1970, I began my postal career as a P/T/F clerk at Ochopee, Florida, Post Office, the smallest Post Office building in the U.S.A. June 1, 1971, I was appointed officer in charge of this office until August 19, 1972, when I was appointed postmaster.* What a happy day! More than 20 years, and I love my work and enjoy arriving at my favorite Post Office each day.

Ochopee is located in the Big Cypress National Preserve in the South Florida Everglades, and is a beautiful, interesting area, although the mosquitoes are a nuisance part of the year.

The Miccosukee Indian Reservation, Seminole Indian camps, and Big Cypress Preserve employees make up a large percentage of the regular customers of mail, which is delivered each day except Sundays and holidays. Tourists are plentiful and make the day a busy one. Throughout Florida, people are told not to miss seeing the smallest Post Office building as they travel the famous Tamiami Trail to Miami and the Florida Keys. It seems not many miss it, as the parking lot fills with cars and tourists with cameras each day. This must be the most photographed Post Office in America. I try to accommodate requests for me to stand in front of this little Post Office for a picture many times a day, and I greet each customer in a friendly way.

Tour buses stop, also, with visitors from all around the world. Purchases of stamps are lots of times in many languages, and the customer and I do not understand each other, but a smile and a little patience overcome that problem.

Over the years, many stories have been done on Ochopee in newspapers, magazines, and television. A TV station from Munich, Germany, visited here, and, with permission from Communications, Tampa, Florida, they did a TV story on this little Post Office. I remember many of the visitors from Germany relating their joy in seeing the program in Germany. Stations in the area use any happening at Ochopee to do another story, such as when the air conditioning was installed in 1987 or when I have an anniversary date each year in October.

The Ochopee Post Office has been a wonderful, big part of my life. I raised two sons, Jack and David, with the help of my husband, Jack. He passed away in November 1986. Many people in the Post Office remember him fondly. He served as president of the Volunteer Fire Fighters' Auxiliary, Florida Branch, for several years, and is the Ochopee punch inventor. The Florida Branch serves Ochopee punch at nearly every national convention, as they will be doing again in Anaheim, California.

I have had membership in the National League of Postmasters since 1971, and have missed very few State and national conventions. It had to be a good reason for me to miss, because Jack always wanted to be there and so did I. I serve on committees at State conventions and help wherever I can. Last year, I served on the Program Committee at National in Atlanta and will serve again this year. I always enjoy being a delegate, and will be one this year in Anaheim.

Editor's Note: Women have always been called "postmaster" and not "postmistress".

I have sincerely tried to be an employee of this business – the United States Postal Service. I have worked hard and have done a good job. May the future allow me a few more years of being postmaster.

Well, I guess I will stop here. I have more than enough clippings to fill many scrapbooks, but I have included a few of my favorites in this book that I will take to Anaheim. I wish I could share these with postmaster friends but never have been able to. I am so happy to have this opportunity to share them with others. May the articles continue in the same fun way and the people continue to visit Ochopee because they appreciate the building and enjoy our friendly service.

"Hokie" (Clara McKay): EVELYN SAVED MY LIFE

Some days I don't do so good, since that alligator took my arm, but I just keep goin'. I tell you, I just loved that Evelyn. I've known her since her son, David, was in diapers. That's been a long time. Evelyn saved my life one day. Years back, when I was takin' care of some old men out back, I cooked and washed for them, and one day I was tryin' to hurry and get my washin' out on the line. I was about to drop dead with a headache. I went and got me a B.C. tablet.

Now, in those days, we got our drinkin' water in jugs from Carnestown. I picked up a jug that I thought was water and took my B.C. It wasn't water at all. It was bleach. I started coughin' and chokin' and screamin'. The only thing I could think of was to call Evelyn at the Post Office. She said, "Hokie, you meet me at the end of your bridge. I'll run over with a gallon of milk and you will drink every bit of it."

I said, "You are runnin' the Post Office. You can't come. Where will you get milk?" Evelyn said, "Don't you worry. I'll run across to the store and meet you at the end of your bridge."

I said, "I'll get you some money for it." She said, "Don't you dare. You do what I say," and she was gone. By the time I got to the end of the bridge, she was there with the milk. She made me drink it. She saved my life that day.

She helped so many people out here besides me. Evelyn was a wonderful girl.

Author Maria Stone with Mama Hokie who lost her right hand to an alligator in at age 81. Mama Hokie had a sign advertising BEER, WORMS outside her café. She is remembered in a song by Gator Nate & the Gladezmen.

Ann Greenwall and Barbara Dwyer:
MEMORIES OF OUR DEAR SISTER

Ann: First, I want to say that, in 1990, we were so proud when my sister, Evelyn Shealy, won the Postmaster of the Year award for the whole State of Florida. She was a celebrity. She won a lot of awards for the Post Office and her work there. She received cards and gifts from all over the world. She went to Anaheim, California, for a Post Office convention to enter a national contest for postmasters. That trip was wonderful, but her health failed after the trip. It might be interesting for people to know that Jack, her husband, made the flagpole at the little Post Office. She was featured in *Peoples* magazine. There was also a big article in the *Naples Daily News* when she snagged a large alligator in Lake Trafford.

When she was not working, she often went on fire calls with the firemen. She was a wonderful person. I am so proud of all that my sister did. I miss her since she is gone. We all miss her.

Barbara: Evelyn was born in Marion, South Carolina, on November 19, 1939, the second of seven daughters. She grew up in the Miami area and lived there until she and Jack moved to Ochopee in the early '60s.

From the mid-'70s to 1982, she and Jack held an annual Thanksgiving "family reunion" on their property for all her sisters and their families and our dad. This was very sweet and generous of them because of the large number of people. There were 30 in attendance at the last reunion. We would all swarm in on them for at least one overnight stay.

Evelyn was a fun-loving optimist. She often said, "What is meant to happen will happen." She never dwelt on problems. She was always outwardly cheerful and was certainly entertaining with all of her tales of her travels, the Post Office, and life in Ochopee.

Evelyn and I took two short trips together – one to Nassau to play black jack in 1987 and another in 1989 to a family reunion in North Carolina. We had such fun, but I especially had the pleasure of getting to really know my big sister.

Although I live in Ft. Myers, 65 miles from Ochopee, Evelyn would think nothing of dropping in on us when she was "in the neighborhood." It was always a nice surprise to see her pop in the kitchen door on her unexpected visits. We really miss her happy chatter and boundless energy. I am so proud of what Evelyn accomplished in her life for her family, friends, and the community.

David Shealy, Evelyn's Son: MY MOM

My mother started at the Ochopee Post Office as a clerk for Mr. Sidney Brown. After he retired, she became postmaster. I was just a little kid.

I remember every morning when Mother opened the big, blue drop box to get the mail out, there would be snakes in there. She would have to get them out before she could retrieve the mail.

Horseflies were a real problem all through the years. Now, there is a sliding screen door which keeps them out. I grew up around the Post Office.

Vince Doerr, Ochopee Fire Chief:
MEMORIES OF EVELYN

I knew Evelyn's husband, Jack, as well as Evelyn. Jack was the first volunteer fire chief in Ochopee. I became Chief when he died. Jack ran the machine shop and campground nearby the Post Office. Evelyn ran the Post Office, but after it was closed for the day, she did all kinds of things to help the volunteer firemen. We parked our fire trucks at their campground. We could work on them at their shop. We had our meetings there under a chickee.

We had about 20 or 30 volunteer firemen. This group was organized in 1974. Evelyn was our treasurer all 18 years that we have been organized. She always brought sandwiches and coffee to us after our meetings.

Because of the lake behind the Post Office, the mosquitoes have always been bad. They seem to swarm especially around the Post Office. I installed a 110-volt air conditioner for her in 1987. She only had a fan all those years.

Evelyn was loved by the whole community.

Now, all the Brazilian pepper trees that grow around the Post Office really protect it from fire, as the pepper trees won't burn. Just the same, I knocked all the growth back quite a ways so now no growth endangers the Post Office during an Everglades fire.

All the nearby area belongs to the Big Cypress National Preserve.

We will always remember Evelyn and her good works at the Post Office and in our community. She did lots of good things for the people who visited the Post Office as well as the people of the community. We were all very proud of Evelyn and our little Post Office. It is the only "brag" we have out here. Everything else is gone.

Ochopee Post Office was awarded an historic marker in 1995 by Collier County.

**United States
Postal Service**

OAKLAND, CA 94615-9996

July 1, 1988

Evelyn Shealy
Postmaster
U.S. Postal Service
Ochopee, FL 33943-9998

Dear Evelyn Shealy:

I am pleased to announce that Ochopee, Florida was smaller than 200 offices entered for the special distinction of "First Place Tie Winner/Smallest Post Office Challenge by Oakland Division Communications."

Thank you for managing one of the nation's three smallest post offices, as determined by the smallest post office challenge. Your "stamp-sized" post office may be small, but it is still very important to us.

I thoroughly agree with Postmaster Shirley Paolini of Bird's Landing, who tied with you for first place, when she said, "Small post offices are good will ambassadors to the public."

Congratulations and keep up the important work you perform for both your community and the Postal Service.

Sincerely,

Linda A. Deaktor
Field Director
Marketing & Communications
Oakland Division

FURTHER READING

WEBSITES with historic documents and pictures:
   Florida State Archives, www.floridamemory.com
   Collier County Museum, www.colliermuseums.com
   Reclaiming the Everglades, http://palm.fcla.edu/
   Cemetery Search, www.findagrave.com

BOOKS with background and recorded interviews:
   Repko, Marya, *A Brief History of the Everglades City Area*
   Tebeau, Charlton W., *Collier County; Florida's Last Frontier*
   Tebeau, Charlton W., *Man in the Everglades*
   Whichello, Jeff, *What Happened to Ochopee?*

PLACES TO VISIT:
   Ochopee Post Office, 38000 Tamiami Trail East (US-41)
   Skunk Ape HQ & Trail Lakes Campground (Shealy Family)
      40904 Tamiami Trail East (US-41), Ochopee, 34141
   Museum of the Everglades. Everglades City, FL
   Collier County Museum, Naples, FL
   Smallwood Store & Museum, Chokoloskee, FL

Editor's Note: This and subsequent pages were added in the 2018 edition.

MARIA'S BOOKS

The End of the Oxcart Trail; the Story of the Roberts Family of Immokalee

The Caxambas Kid;
    The Life & Times of Famous Fishing Guide Preston Sawyer

Dwellers of Sawgrass and Sand, volume I
Dwellers of Sawgrass and Sand, volume II
Dwellers of Sawgrass and Sand, volume III

The Good Ole' Days in Naples and Collier County

Ochopee; The Story of the Smallest Post Office

Old Soldiers Have Nine Lives

Swamp Buggy Fever

The Tamiami Trail; A Collection of Stories

We Also Came; Black People in Collier County

Maria and her husband Peter in 1992 at the unveiling of the Tamiami Trail signs.
Maria died in 2009 at age 86 in Naples where she had been given the Key to the City.
Peter was devoted to Maria and published all her books but he passed away before her.

Collier County Stamp Club presents

NAPLEX '92'

March 21, 22

HONORING
The Tiny Post Office
at OCHOPEE FL
- * -
Old Naples Depot
Naples FL

THE 1992 SPECIAL CANCEL

The pictorial cancel of the tiny Post Office at Ochopee, Florida was arranged for through the generous cooperation of Postmaster Thomas Sapp and the U.S. Postal Service.

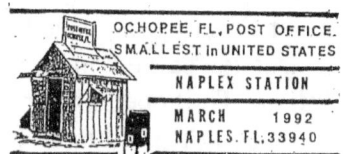

The design depicts the Post Office as it appears today on the Tamiami Trail (U.S.41) 35 miles east of Naples, Florida.

The cancel will be applied to all mail mailed at the show on March 21 &22 1992.

* * *

The U.S. Postal Service 'Naplex' Station provides an opportunity for collectors and the public to purchase stamps and postal material not always available.

The Post Office Postique, located at the Main Post Office on Goodlette Rd. is one of a few in Florida with special services for collectors, stocking new issues shortly after their issue.

Ask for Dave Pittlecow, at the Philatelic window for new issues and other postal items.

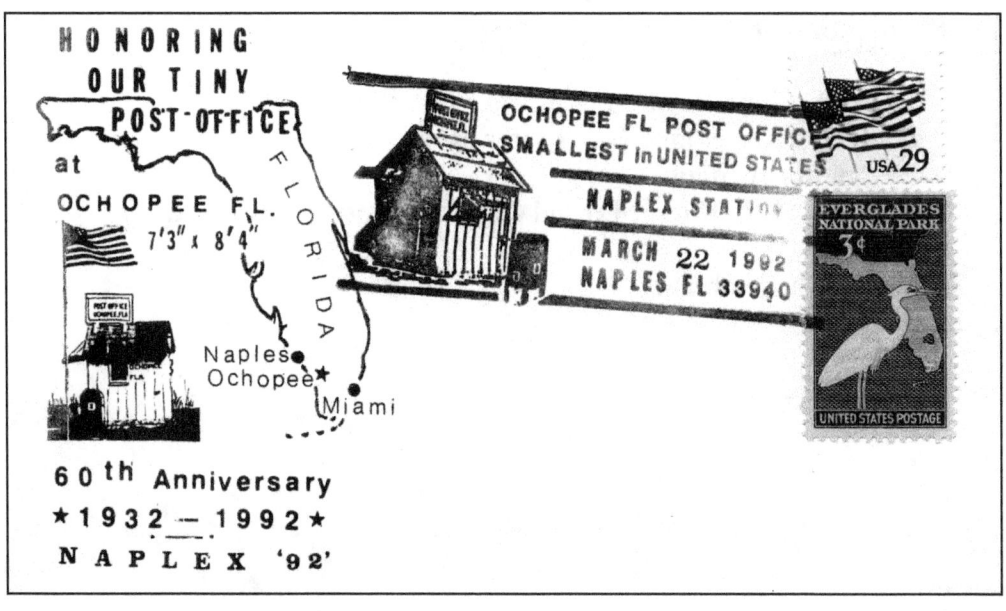

TIME LINE

1873 James Franklin Jaudon born on October 19 in Waco, TX
1900 James Tate Gaunt born on November 8 in Miami, FL
1915 Jaudon began Tamiami Trail from Miami
1917 United States entered World War I
1919 United States passed 18th Amendment enabling Prohibition
1919 United States passed 19th Amendment enabling votes for women

1923 Collier County established by Florida legislature on May 8
1923 construction started on Tamiami Trail and town of Everglades
1928 Tamiami Trail opening ceremony on April 26 in town of Everglades
1928 Gaunt bought land in Ochopee, started farming
1929 Great Depression started with stock market crash on October 29
1932 Post Office in Ochopee licensed
1938 Jaudon died on February 22 in Miami

1953 Ochopee store with Post Office burnt down on May 12
1960 Hurricane Donna struck the southwest coast of Florida
1962 Gaunt moved to LaBelle, Florida
1963 Ochopee Post Office moved back during Trail widening
1965 Ochopee tomato farm closed

1971 Evelyn Shealy started working at Ochopee Post Office
1972 Gaunt retired from farming
1974 Big Cypress National Preserve established
1978 Ochopee Post Office listed on National Register of Historic Places
1978 Gaunt honored by LCEC

1987 James T. Gaunt died on December 20
1990 Evelyn Shealy named Postmaster of the Year
1991 Evelyn Shealy died
1995 Ochopee Post Office awarded Collier County historic marker

2016 Monroe Station burnt down

www.ingramcontent.com/pod-product-compliance
Lightning Source LLC
Chambersburg PA
CBHW060519300426
44112CB00017B/2734